BASICS OF OCCUPATIONAL SAFETY AND HEALTH

THOMAS P. FULLER

nsc
PRESS
National Safety Council

NSC Press Editor: Deborah Meyer
Cover and interior design: Jen Villarreal
Senior Director, Publications: Suzanne Powills
Cover Photos: trabachar/iStock/Thinkstock; sirastock/iStock/Thinkstock

NATIONAL SAFETY COUNCIL MISSION STATEMENT

The mission of the National Safety Council is to educate and influence society to adopt safety, health, and environmental policies, practices, and procedures that prevent and mitigate human suffering and economic losses arising from preventable causes.

DISCLAIMER

Although the information and recommendations contained in this publication have been compiled from sources believed to be reliable, the National Safety Council makes no guarantee as to, and assumes no responsibility for, the correctness, sufficiency, or completeness of such information or recommendations. Other or additional safety measures may be required under particular circumstances.

Printed in the United States of America

Library of Congress Cataloging-in-Publication Data
Fuller, Thomas P., author.
Basics of occupational safety and health / Thomas P. Fuller.
pages cm
Includes index.
ISBN 978-0-87912-333-8
1. Industrial safety. 2. Industrial hygiene. I. Title.
T55.F85 2014
363.11--dc23
2014016726

ISBN: 978-0-87912-333-8
NSC Press Product Number: 176450000

ABOUT THE AUTHOR

Thomas P. Fuller, ScD, CIH, MSPH, MBA, is the director and assistant professor of the safety program at Illinois State University. He teaches courses on the safety profession, safety technology, regulatory affairs, hazardous materials management, and toxicology. He earned his doctorate in industrial hygiene at the University of Massachusetts Lowell. Dr. Fuller also has a Master of Science in public health from the University of North Carolina, and a master's in business administration from Suffolk University. He has 34 years of experience working in occupational health and safety in a variety of industries, including hospitals, nuclear power plants, universities, pharmaceuticals, and manufacturing.

ACKNOWLEDGMENT

This revised and updated edition of *Basics of Safety and Health* is based on the text originally revised by Neville C. Tompkins. Thank you to Mr. Tompkins for providing such a broad and durable safety and health fundamentals foundation to build upon.

CONTENTS

PREFACE

Occupational safety and health play a key role in the success of any organization. The ability to reduce the probability and severity of accidents that cost time and money can be a way for an organization to differentiate itself from competitors and to improve its efficiencies. A safer workplace can help ensure that your highly skilled workers will not be hurt or disabled and need to be replaced. In addition, making the workplace safe for employees reduces the possibility of damage to expensive equipment and systems that then leads to lost production capabilities and higher insurance premiums. An effective safety culture will lower the potential for costly regulatory enforcement, improve employee morale and productivity, and increase the quality of the organization's services or products.

Readers of this book will be able to quickly learn about the most important aspects of building and running a modern safety program. Fundamental principles and basic definitions are presented in a straightforward, conversational manner. *Basics of Occupational Safety and Health* provides a broad overview of many of the concepts and principles used to make a case for having a good safety program and for getting the support of both workers and management.

This book is useful to a broad range of professionals with diverse skill sets. People with experience and education in other areas and who now find themselves responsible for some aspect of occupational safety and health can use this book as a starting point to gain a basic awareness of safety issues. The security officer who is now in charge of responding to regulatory agency citations for poor workplace protections could be one example. The director of human resources who is already responsible for workers' compensation claims may suddenly be tasked with recording and reporting injuries and illnesses for regulatory agencies. Senior business managers can use this book to get a skyline view of safety and health, to find out why safety and health are important to the success of their organization, and to find out why the safety program needs their attention and support.

Most importantly, this book is about protecting employees. A good safety program that is fully implemented and continuously improving is the best way to ensure that everyone who comes to work goes home each day. A culture of safety is one where workers understand that the safety program has their best interests at heart and that they help develop the program and are part of its continuous improvement. Employees and their representatives can read this book to become aware of what a good safety program looks like and to recognize where the safety program in their workplace may be deficient.

This book can also be used as a text for an introductory university course on occupational safety and health. It is easy to read, and it can be completed in one semester to supplement other readings and activities. It is a great way for professors to introduce their students to the safety and health industry and profession. As a supplementary teaching tool, end-of-chapter questions and their answers are available on the National Safety Council Faculty Portal.

This book is based on Neville Tompkins' revision of the *Basics of Safety and Health*; however, it has been updated to include new information on the anticipation and recognition of hazards, with a focus on occupational safety and health. A new chapter on hazard evaluation and risk assessment has also been added. Plus, this new edition highlights the changes in regulatory programs and rules within the occupational safety and health field.

1

WHY ARE SAFETY AND HEALTH IMPORTANT?

Whether you are a manager, a front-line worker, a head of human resources, or an employee representative, the health and well-being of all employees is important to you and vital to the success of the organization. Everyone in the company plays a role in safety, and everyone should know a little bit about what it takes to have an effective safety program.

If you are in management, you should have a handle on what safety programs are necessary and who in the company is running these programs. You also need to know whether these programs are doing what they're intended to do: ensure a safe work environment. Workers and their representatives should have a clear understanding of their legal rights to a safe workplace and a basic understanding of the features of a safe workplace.

If you have been tasked with taking on the occupational safety and health program and you don't have formal training or prior experience in the area, this book is a great introduction to the field.

While you inherently know that safety is "good" and injuries are "bad," you may be feeling apprehensive about overseeing an area of which you have little knowledge. You may not possess the educational background or the work experience required to organize and implement a safety management process.

This book is intended as an orientation to occupational safety and health for students, nonsafety professionals, managers, and workers. It provides guidance and suggested techniques that have been used effectively at dozens of locations across many industries and that have application to almost any workplace where some form of job hazard exists.

A definition

Simply put, occupational safety is concerned with identifying, assessing, and eliminating or controlling workplace hazards.

How big is the problem? A total of 4,383 U.S. workers died of unintentional workplace injuries in 2012, according to the Bureau of Labor and Statistics. Another 3.7 million suffered disabling injuries on the job, according to the National Safety Council. While the human toll of these occupational injuries and deaths is huge, the costs associated with workplace deaths and injuries also can adversely affect any company's bottom line. In 2012, the average cost of a worker death was $1.39 million, and the cost of each disabling injury was $54,000, according to National Safety Council statistics.

Workers' compensation costs also are growing. Occupational injuries and illnesses cost American employers more than $53 billion per year. Indirect costs—such as lost productivity, new employee training and replacement costs, and administrative costs associated with accident and injury investigation and support—can top $100 billion per year. Losses to workers and their families in lost wages and the costs of caring for the injured are other monetary losses that result from inadequate worker safety.

In fact, the economic effect of unsafe practices is one reason why more companies are focusing on safety. Other companies cite ethical reasons, stating that providing a safe workplace for employees is "the right thing to do." Some companies begin looking into safety as a result of a government inspection, citation, or penalty. Still others may consider the negative media publicity or an unsatisfactory reputation as "not a good place to work" as reason enough to begin paying attention to safety.

These considerations all lead to needing formal safety and health processes that identify and control workplace hazards and unsafe practices that lead to incidents.

Safety and the law

American employers have a legal obligation to "furnish to each of [their] employees employment and a place of employment which are free from recognized hazards that are causing or are likely to cause death or serious physical harm to [their] employees." That is not a vague statement of intent but the specific federal law outlined in paragraph 5(a) of the Occupational Safety and Health Act of 1970, which is enforced by the Occupational Safety and

Health Administration (OSHA), the federal agency tasked with overseeing and enforcing safety and health in U.S. businesses.

Approximately 22 "Agreement" states have assumed OSHA's responsibilities for promulgation and enforcement of their own regulations for worker safety. Sometimes these states implement stricter safety requirements than the federal regulations mandate. Additionally, some counties have legislated requirements for personal safety and for transportation and road safety. And some cities and towns have enacted requirements regulating work conduct in plants, on construction sites, and in offices within their municipalities.

Other federal agencies and laws also have occupational safety and health implications for businesses. The Environmental Protection Agency, for example, enforces environmental regulations that directly impact worker safety. Pollution from business operations on land, surface water, ground water, and air can result in the exposure of people, plants, and animals to harmful substances or energy sources. Other examples: The Federal Communications Commission (FCC) sets standards and limits on occupational exposure to radio waves and communications radiation emissions. The Mine Safety and Health Administration (MSHA) regulates worker safety and health in the mining industry.

Among the other laws governing safety are the Rehabilitation Act of 1973 and the Americans with Disabilities Act, both covering the safety of disabled employees; the Fair Labor Standards Act, covering hazardous occupations for working persons under age 18; the National Labor Relations Act, covering work conditions; and Title VII of the Civil Rights Act, covering discriminatory treatment in work assignments.

Lack of compliance with these regulations can result in citations, fines, criminal charges, and unfavorable publicity for the unwary employer.

The potential for criminal charges

Safety is serious business. A workplace fatality, an explosion or fire, or injuries to several employees will quickly receive media attention. Within hours, an OSHA compliance officer will be on site to find out what happened and determine if any OSHA regulations were violated.

An employer convicted of a willful violation of a standard that resulted in the death of an employee faces a court-imposed fine, imprisonment of up to six months, or both. Under the Sentencing Reform Act of 1984, a criminal conviction could result in fines of up to $250,000 for an individual and up to $500,000 for a corporation. Further, criminal charges can be filed in an incident resulting from management or supervisory misconduct. Some prosecutors are using charges such as *manslaughter* or *reckless conduct* in cases of workplace deaths and *assault and battery* in cases of serious workplace injuries. These prosecutors go beyond the provisions of the Occupational Safety and Health Act and related environmental statutes to indict and, in some cases, convict companies and their executives under criminal or civil laws for workplace safety and health violations.

Making safety part of the business plan

Despite the rising costs of workplace injuries and the potential legal pitfalls, it is still difficult for safety to gain upper management's attention. Instead, management's attention often is consumed by the demands of operating a business in today's economy, and thus many organizations give safety very little time, attention, or budget.

This is unfortunate. Executive management that does not embrace safety, bringing it into the mainstream of business activities, sends the message that safety requires less attention from key managers, who ultimately are responsible for the success of incident prevention.

Conversely, successful companies have incorporated safety-related activities into business operations, meetings, and planning conferences. These companies place safety and health on the same level as efficiency, quality, and cost control. They know that incident prevention is as important as any other aspect of business operations and should be handled in a manner similar to dealing with machine breakdowns, efficiency losses, personnel turnover, job absences, and quality problems. The successful company has a culture that promotes safety and rewards safe practices.

Partnering with employees

Who should be interested in safety? Everyone within a company.

- Workers—whose safety is in jeopardy in an unsafe workplace—should care about their personal safety.
- Supervisors should be interested in the safety of workers in their departments. Some supervisors' compensation may be based in part on controlling injuries and enhancing safe operating practices.
- Management should be interested in the bottom-line effect of safety, including employee morale and employee turnover, workers' compensation and

health care costs, production inefficiencies and reduced product quality, and potential OSHA penalties and negative publicity. Factors affecting safety—such as design of machine guards, budget items for building and equipment improvements, priorities for engineering improvements, trainer selection for safety sessions, and the creation of a culture of safety—are all under management's control.

- Labor organizations represent work safety and interests in a coordinated and informed manner. These groups often have safety professionals on staff who can identify workplace hazards and work with management to ensure that controls to reduce those hazards are implemented.

Safety is not a single person's or department's responsibility. In fact, spreading this responsibility to everyone and clearly communicating that responsibility will generate greater involvement. Although management has the ultimate responsibility for workplace safety, managers and supervisors should not have to shoulder that responsibility alone. Indeed, the responsibility can be shared with employees through their meaningful involvement in the safety process. In fact, no workplace activity lends itself better to employee involvement than safety.

Employees may know more about the hazards and risks of their jobs than anyone else; after all, their personal safety is at stake. Those who are empowered and trained to assist in safety will sharpen their safety problem-solving skills, become hazard detectives, and evolve into more safety-conscious workers. The results are employees who take ownership of safety and, in the end, fewer workplace injuries.

Forward-thinking employers involve workers in safety and health activities such as safety committees, facility safety inspections, and incident investigations. Such involvement promises much greater long-term success in safety than when management tries to supervise the safety process alone. Continuous improvement and going beyond minimum safety requirements are integral parts of an effective safety culture.

Using job performance as an incentive

One way to raise safety's importance throughout the company is to build safety measures into annual performance evaluations and merit reviews. If safety and health performance is included as a factor in the determination of a salary increase, wage improvement, or bonus payment, it will automatically heighten the importance of safety.

Most companies hold employees at all levels accountable for performance in terms of productivity, quality, or cost control. And although the real reward of working safely is not getting hurt, performance achievements should go beyond that to include incentives paid when incident prevention is working and incentives not paid or withheld if hazards go undetected or uncontrolled and workers are injured.

Success is achieved when management realizes that the same skills that resulted in outstanding production, quality, and cost controls can deliver similar results for safety and health.

Avoiding the "quick fix"

What does it take for an organization to succeed in safety? Some businesses believe the answer is a "quick fix" prompted by a serious incident, an OSHA citation, or runaway workers' compensation costs. The quick fix may consist of new warning signs or flashy banners placed near the entrance proclaiming, "Safety First." In time, however, management returns to regular business, interest in safety wanes, the posters and banners get dusty, and the quick fix quickly starts to fizzle.

This book references "safety processes" with an emphasis on weaving hazard anticipation, recognition, evaluation, and control into a safety process that becomes part of the core activities of the business or organization, along with cost control, manufacturing efficiency, improved customer service, and quality control.

Safety does pay. And integrating safe work practices and hazard detection and control into the way an organization does business is the way it pays because it reduces costs related to injuries, improves efficiencies, and increases the value of the business. It must start at the top and involve all managers, supervisors, and employees. Upper management that understands the business value of effective safety and loss control activities will quickly gain support from managers and supervisors at lower levels. True leadership and commitment will help eliminate obstacles, lend credibility, establish priorities, and provide the capability to get things done.

Coming next ...

The next chapter will show that safety should not be a stand-alone activity. In fact, many other business activities can be used as building blocks to an effective safety and health process. But how do you secure senior management's commitment?

2

BUILDING ON WHAT YOU HAVE

Safety is not a stand-alone activity. In fact, it may not succeed if management sets up safety as an isolated activity within a business. Safety and health should instead be incorporated into and throughout all business operations. And while this statement may seem overwhelming to those new to safety, starting from scratch may not be necessary. Most organizations already have some activities that can be used as a starting point for additional initiatives to detect and correct hazards. Many of these activities are conducted under the auspices of "maintenance." The following are three examples of safety activities that may already be conducted in a business.

1. An electrical supervisor and an electrician conduct a monthly inspection of a company's electrical control boxes. Although this activity is considered a maintenance procedure because problems are corrected promptly, it also is a safety activity.
2. A staff engineer conducts a monthly review of fire extinguishers, valves, hoses, and alarm systems. These are all ingredients of an effective fire protection and emergency evacuation plan. This routine review is part of a working incident prevention process with obvious, positive injury prevention implications.
3. A human resources recruiter carefully checks the work backgrounds and business references of job candidates. This activity also ensures that the information on candidates' employment applications—such as machine operation capabilities and experience—is factual. The implication for safety is that the candidate is experienced and will be less likely to be hurt on the job.

Getting started

Any existing, informal incident prevention activities that are working should first be reviewed and assessed for effectiveness as well as possible improvements and/or eliminations.

Again, it is much easier to build on an activity that is working well and is already in the mainstream of the business than to start from scratch. Sometimes all that is needed is a periodic review of existing programs to ensure that there is continuous improvement and that any new regulations have been incorporated into systems and programs.

Safety program* versus *safety process

Safety program	*Safety process*
Focused on injuries	Long-term and continuously improving
Employee-driven	Led by management with employee involvement
Reactive to injuries only	Proactively identifies and controls incidents before they occur
Applied to hourly employees only	Developed with employee input
Predetermined program elements	Flexible to address hazards
Disciplinary actions taken against those working unsafely	Incorporated into job performance
Unsafe work conditions reported after an injury	Hazard detection process spots unsafe practices and procedures before problems occur
Compliance focused	Extends beyond compliance

So what makes a successful safety process? A hazard detection, elimination, or control activity should be an integrated process no different from building code inspections, engineering reviews, machinery inspections, employee communications activities, and tours by insurance loss prevention specialists. Hazard detection and control activities shouldn't start and stop but, rather, should be part of a continuing process of detection, assessment, elimination, or control of potential injuries and illnesses.

Like a total quality process, the success of a safety process is based on continuing and ever-evolving efforts to improve. When integrated into the company structure, hazard detection and control become integral parts of business, requiring the same attention and resources from senior management as any other component of business operations.

Still, many senior managers don't consider the business implications of an inadequate safety process. For example, a sales drive to increase order size from smaller customers requires stopping and starting machines frequently and, therefore, increases the opportunity for work injury.

For this reason, it is important to communicate the positive effects of injury reduction, such as lower workers' compensation costs, reduced absences from work, minimized legal vulnerability because of effective compliance programs, and improved productivity and worker morale.

However, terms such as *incident frequency* and *severity rate* may not be meaningful to business leaders unless those leaders understand the impact of those components on the bottom line. Safety benefits need to be translated into business benefits. For example, a safety process can improve cost control and efficiency. Fewer work injuries can lower the costs of hiring and training replacements. The bottom line is that work injuries drain profits.

In addition, most company management decisions are based on a cost-benefit analysis. Will the purchase/improvement/change result in reduced costs, improved customer service, or better quality? For any proposed change in process, policy, or equipment, management considers what will be gained by the change versus the cost of the change. Benefits might include reduced or eliminated expenses, increased productivity, or better quality.

The cost of safety

Knowing both the direct and the indirect costs of injuries is important. Direct costs are fairly evident. They include **medical costs, compensation paid to an injured employee, and cost to repair or replace damaged buildings, equipment, materials, or products.**

Indirect costs are not as obvious. Such costs include:

- Time lost from work by the injured employee
- Employee's loss of earning power
- Economic loss to injured worker's family
- Cost of finding and hiring replacement employees
- Administrative costs associated with report preparation, insurance claims,

and investigations
- Supervisory time distraction
- Cost of training replacement employees
- Cost of damaged equipment not in service
- Spoiled product or interrupted service
- Loss of production and profit from idle machines and processes
- Delay in filling customer orders and late fees or penalties associated with delays

Calculating the direct and indirect costs of an injury

Direct cost

Enter the total value of a workers' compensation claim for a specific injury or a job illness, including medical costs and temporary total disability payments.

Direct cost: $________

Indirect cost

Multiply the total direct cost from above by the appropriate cost multiplier below.

If the direct cost is	*Use this cost multiplier*
Up to $2,999	4.5
$3,000 to $4,999	1.6
$5,000 to $9,999	1.2
$10,000 or more	1.1

Direct cost × cost multiplier = $____________
Indirect cost: $____________
Direct cost + indirect cost = total cost
Total cost of the job injury or illness: $__________

Cost and incident data also can be compared internally among departments or among locations as well as externally among other, similar companies. External data may be obtained from OSHA, which issues an annual summary of occupational injury and illness rates for the previous fiscal year. This national information is broken down further by industry, allowing employers to compare their own incident statistics to the summary of those for similar companies in the same industry.

Leadership support also can be obtained by showing the cause and effect of safety on company values. In some organizations, the value is "efficiency"; in others, it is "quality." Some companies tout "the best in customer service." The American Industrial Hygiene Association has recently developed an approach called the "Business Value Strategy" to workplace safety and health. Whatever the term, managers and employees are invited to rally around these slogans and support these values.

These values also carry safety implications. Being best in customer service cannot be achieved when key employees are absent due to work injuries. Sound operating principles include efficient operations and good morale, which are heightened by employees working effectively and safely without fear of injury. For the efficient profit-oriented company, the immediate and long-term effects of hazard detection and control are fewer injuries and job disruptions and lower workers' compensation costs.

A safety and health policy

A policy statement from a ranking manager can have a positive effect at the outset of a new or reinvigorated hazard detection and control activity. The following is a template for a safety and health policy statement.

To: All staff
Subject: Our commitment to a safe and healthy workplace

This organization is committed to a safe and healthful workplace where injuries, illnesses, property damage, and environmental impacts are minimized to the greatest extent possible. This commitment stems from the belief that a safe, healthy, and environmentally friendly operation will also be the most productive and efficient. The health of the organization depends on the health of its employees.

Management, supervisors, and workers all share a responsibility for safety and health. Safety and health roles and responsibilities have been delineated in policies, programs, and procedures. Everyone is expected to receive required safety training and to follow required procedures at all times.

When accidents or incidents occur or other areas needing improvement become apparent, all employees are encouraged to report the events and to notify responsible individuals of the ideas for improvement. A culture of safety, trust, and continuous improvement is the foundation of a safe and healthy workplace.

Finally, executives like moments of pride. Being able to tout that they have one of the safest work environments in their industry will raise enthusiasm among employees, their families, the community, and the industry. Some trade and industry associations hand out awards to the best achievers. If a company is at the bottom of the pile in its industry, the opportunity to make a case for safety improvement is at its peak. If the organization is at the top, the case should be made to stay there.

Whose job is it?

A production manager is responsible for production efficiency, quality, and costs. The information technology department manager is responsible for quality of output, computer equipment conditions, and worker competence. The line function has the power and authority to implement—and is held accountable for—quality, cost, and production parameters. And staff functions—such as accounting, human resources, and scheduling—have the specialized knowledge to assist the other departments.

So it stands to reason that the safety coordinator should be responsible for identifying, assessing, and controlling or eliminating hazards throughout an organization—right? Wrong!

Everyone—from top management down to each employee—is responsible for safe working conditions and safe operating practices. In addition, all are responsible for ensuring that equipment is well maintained and operated safely, that workers are trained in the safe operation of the equipment, and that a follow-up or hazard detection and control system eliminates or controls existing hazards and quickly detects new hazards.

Some people are more likely to agree to safe operating procedures and work conditions than others. A human resources manager with safety responsibility or other department managers, supervisors, or line managers who know the importance of safety will be more likely to support the safety process.

Managers should regularly tour business facilities—observing activities, encouraging positive behaviors, and correcting unsafe activities. An easy and natural part of these key managers' conversations can be commending an employee's safe behavior, inquiring about the status of an incident victim, or thanking safety committee members for their efforts.

Keeping everyone motivated and excited about safety doesn't have to be difficult. A few words about safety from key executives can go a long way toward enhancing safety efforts. Key managers can be a powerful example and encourage positive behaviors by wearing protective gear appropriate to the particular workplace.

Management should widely publicize safety meeting minutes, the agenda of upcoming safety meetings, and safety successes and let people know that positive things are happening on the safety scene. Management should also give opening comments about safety at a business meeting, add an addendum regarding safety to operating statements, or honor supervisors for safety leadership in one of their meetings.

Often management will say they are "100 percent behind safety," and they mean it. But remember that management has many factors pressing for attention—cost controls, customer service, quality, personnel, and equipment upgrades. Nevertheless, hazard detection and control merit a fair share of management time, support, and budgetary allocation for positive safety activities.

Also remember that hazard detection and control never cease. Initial surveys and control measures can do a great deal to put basic steps and procedures in place, but hazard detection and control are high-level, continuing management responsibilities.

Safety as a culture

Once embedded into daily business activities, the safety and health processes become part of the corporate culture of a business. An organization's culture is composed of its values, beliefs, mission, goals, performance measures, and sense of responsibility to its employees, customers, and communities in which it operates. These values and beliefs are translated into a system of expected behaviors, areas of emphasis, policies, procedures, and day-to-day details of operations. These behaviors and areas include:

- Commitment (or noncommitment) to injury prevention
- Management involvement in and direction of positive safety activities
- Managers, supervisors, and workers being responsible for hazard detection and control
- A supportive structure of assistance and resources
- Safety not being sacrificed as a "cost-cutting" measure
- The safety program growing with the company and industry, when required
- Safety standards and methods incorporated into standard operating procedures
- Continuous improvement in all phases of business activity, including quality, cost control, delivery, performance, and injury prevention

Coming next ...

The next chapter will begin to describe how workplace hazards are anticipated and recognized: through a broad and methodical thought process and way of thinking about possible workplace hazards.

3

ANTICIPATION AND RECOGNITION OF HAZARDS

Hazards are a part of every industry and every workplace. The purpose of safety is to find hazards—such as dangerous workplace conditions or unsafe work practices—and eliminate or control them before they cause injuries.

Said another way, a hazard that is not identified, eliminated, or controlled may result in injury or damage to equipment, buildings, materials, or people. And some degree of hazard can be found in every business or service.

In fact, hazard identification and control are at the core of a successful injury prevention activity. The following steps provide the basis for identifying and controlling hazards.

- Step 1. Study the industry and similar businesses to anticipate hazards.
- Step 2. Outline specific business processes, systems, raw materials, by-products, and products.
- Step 3. Identify the most hazardous materials, processes, or work areas for further assessment.
- Step 4. Evaluate and qualitatively or quantitatively measure risk.
- Step 5. Identify where hazardous processes or materials can be eliminated or substituted.
- Step 6. Implement physical engineering controls to separate and protect workers from hazards.
- Step 7. Create administrative controls to reduce exposures to hazards.
- Step 8. Identify and supply personal protective equipment to reduce risks.

Data regarding injury rates and statistics are available for a broad range of businesses from a variety of sources. The Bureau of Labor Statistics publishes annual reports of workplace injuries and fatalities, broken down into numerous industrial and occupational sectors, which makes it easy to identify the most likely hazards in any workplace.

Professional organizations such as the American Industrial Hygiene Association (AIHA) and the American Society of Safety Engineers (ASSE) can also be consulted to find safety information about workplaces. The Occupational Safety and Health Administration (OSHA) and the National Institute for Occupational Safety and Health (NIOSH) are two federal government organizations that provide a wide range of information on the hazards associated with numerous industries and workplaces. Information on the latest statistics and trends can be found on their organizational websites. Some of these are shown in Figure 3.1.

Occupational Safety and Health Administration
www.OSHA.gov

National Institute of Occupational Safety and Health
www.NIOSH.gov

American Industrial Hygiene Association
www.AIHA.org

American Society of Safety Engineers
www.asse.org

National Safety Council
www.nsc.org

Figure 3.1. Links to Safety and Health Information and Data Sources

Additional sources of information on a given industry include peer-reviewed scientific literature and studies, government reports from other countries, and published analyses from other, similar industries.

Once external sources of information about an organization have been reviewed, the company's database of injury and illness becomes a key source of safety data. Most organizations keep basic documents that can provide clues to the business's inherent hazards. Examples of such documents include first aid records, OSHA logs, workers' compensation reports, and OSHA citations. Data from these documents can help pinpoint safety hazards.

Experience modification rates (EMRs)

A useful piece of data found within workers' compensation reports is the experience modification rate. This rate is used by insurance organizations to determine workers' compensation premiums using a formula developed by the National Council on Compensation Insurance, a trade association for workers' compensation administrators.

Using the formula, an insurance carrier estimates the financial losses expected during a period of time—typically three years—and later compares the estimate to the actual losses encountered by a company based on the claims against the insurance policy. The formula accounts for company size, type of business, and other employer-specific factors.

Each company begins with an EMR of 1.00. At this rate, a company must pay 100 percent of the average workers' compensation premium for its class of work. If the rate is below 1.00, the premium is reduced; if it is greater than 1.00, the premium is increased. If the company's adjusted EMR is greater than 1.00, safety efforts are less effective than the average; if less than 1.00, injury prevention efforts are more effective than average. This rate is another indicator of the cost-effectiveness of hazard identification and control.

First aid log. Review of the first aid log should provide a broad picture of the types of injuries and illnesses encountered at the workplace. First aid logs usually list every injury requiring first aid, including simple finger cuts, sore arms and legs, headaches, and injuries resulting from slips or falls. A review of the first aid log, for at least one year prior, is the starting point to determine where injuries are occurring.

Workers' compensation reports. A review should be conducted of individual claim forms as well as of the insurance carrier's summary of losses from active and unresolved workers' compensation cases. Workers' compensation "First Report of Injury" forms offer details of workplace incidents that required outside medical treatment or that may have involved extended rehabilitation services or disability.

The OSHA 300 Log. The OSHA 300 Log is used to document job injuries and illnesses for employers of more than 10 people, as required by OSHA. The log records injuries requiring medical treatment beyond first aid as well as any lost workdays resulting from injury. The entries on the log may parallel the cases listed on the company workers' compensation report filings.

Calculating incidence rates

The OSHA 300 Log is a good tool for calculating a company's incidence rate—the frequency of serious incidents. The incidence rate uses a formula based on the premise that the average employee works 2,000 hours per year—or 40 hours × 50 weeks. The rate is based on 100 employees. Therefore, 100 employees working 2,000 hours each is 200,000 total work hours. The incidence rate formula is:

Number of injuries × 200,000 ÷ employee hours worked per year = incidence rate

Injury severity rates or lost workday incidence rates also can be calculated using the same formula.

Number of lost workdays due to injury × 200,000 ÷ employee hours worked = lost workday incidence rate

Employers can compare their injury rates to the rates in similar industries published annually by the Bureau of Labor Statistics.

OSHA inspections. Reviewing OSHA inspection files is an important step in determining the degree of danger existing at a workplace. There may be some correlation between the inspection summary and trends in the first aid log, OSHA 300 Log, or workers' compensation reports.

A review of this type of data should point to causes of injuries. With these specific facts in hand, management is in a much better position to focus its safety and health efforts on problem machinery, conditions, specific jobs, or work practices that are causing injuries.

Armed with factual information about the most common types of injuries and illnesses in a given work setting, it is possible to anticipate what types of hazards may be found at any specific work setting. Before stepping foot in a hospital, for example, one might imagine occupational exposure to hazardous infectious agents in the workplace. Exposure to chemical disinfection and

Example of data review

The following is an example of the results revealed by a small manufacturing company's review of its first aid log, OSHA 300 Log, and incident reports.

The company found that although no OSHA inspections occurred over the past three years, OSHA recordable cases did occur. These cases were spread across most departments and showed no apparent trends. The company also found that very few injuries occurred in the electrical shop, where trained electricians work.

However, the data also identified the following areas needing further investigation:

1. Most first aid cases resulted from hand, arm, and wrist cuts in the subassembly, press forging, and maintenance service departments.
2. One finger amputation occurred in the press forging department—a single but serious injury.
3. First aid cases showed many types of injuries in the hand-packing department.
4. Employees in the hand-assembly department complained about sore arms and backs.
5. Strain injuries were more likely among summer vacation replacement hires.
6. Despite improved ventilation and workers wearing face masks, workers in the chemical-mixing area complained of headaches.
7. Three tripping injuries occurred in the packing department resulting from poor housekeeping.

The data also pinpointed hazards:

- In the subassembly, press forging, maintenance service, hand-packing, and chemical-mixing departments
- Among summer replacement hires
- To the upper extremities
- Consisting of cuts, strains, and headaches

cleaning agents would also be anticipated. From available information and statistical databases, you would be aware that musculoskeletal injuries associated with patient handling are one of the leading causes of occupational injury and lost workdays in hospitals. You could, therefore, anticipate that the hospital you were entering would have similar hazards.

Even without formal safety or industrial hygiene training, by thinking outside the box and imagining what possible work activities are performed and what adverse consequences could occur, it is possible to begin to anticipate what hazards might be expected and what to look for more closely. In new industries and technologies with very little historical or operational data, it is prudent to consider the "worst-case" scenarios. A brand-new chemical that is believed to be "perfectly safe" should be handled with the greatest of care until all the required toxicological testing data are in hand. History is full of new materials that were considered perfectly safe until the health effects and disastrous properties became known years later. Radioactive materials are perhaps the best example of this; Marie Curie and Henri Becquerel, who discovered radioactivity, are just two of the earliest researchers in this field who died of radiation poisoning or related diseases.

In some cases, such as in a new industry or technology, it is necessary to learn more about the processes, materials, and products before it is truly possible to anticipate the hazards. Information on the general industry and specific companies making or producing the product can provide some insight into the possible hazards. Published peer-reviewed technical papers can tell us a lot about the methods and materials used to produce a product. This information can begin to enlighten a trained occupational safety and health specialist about potential hazards.

By learning about the new technologies and processes, the safety and health professional can more adequately assess the hazards and risks associated with different jobs and work areas. Assistance may be available from professional organizations that specialize in certain industries. Consultants with expertise in special industries may also be called in for additional direction and support when necessary.

For example, without knowing anything about 3D printing, it is possible to gather enough information to learn that it often involves spraying aerosolized metals or plastics in a series of layers to form a three-dimensional product. Without even knowing which chemicals are used, it is possible to anticipate workers' exposure to the chemical through inhalation during the spraying process or dermal exposure through handling the material in setup or cleanup of the work space.

By fully anticipating potential hazards in a workplace, it is possible to more accurately recognize the hazards that actually occur in each workplace setting. Recognition involves a more thorough identification of each hazard in each particular work setting and of the various steps and tasks in the work process.

In the recognition phase of hazard assessment, the processes, chemicals, and other workplace hazards are listed for each of the various work process steps, tasks, and work areas. A comprehensive overview of the factory or institution is developed. The quantities of various chemicals are listed, along with the types of handling practices and uses at each step/task/location. Safety Data

Sheets may be obtained for various products and chemicals used in the process or during workplace activities. For example, if we anticipated that hospital workers were exposed to hazardous sterilants, then in recognition we would identify which areas of the hospital use sterilants, how much they have, what they do with them, and exactly which types they use.

Recognition can be a daunting task at any facility, and it should not be underemphasized or overlooked. Some of the most hazardous chemicals may be kept in small quantities in limited areas. Activities that are conducted only infrequently can still present high levels of occupational risk due to the extreme severity and consequences of exposure.

Other safety hazards are more difficult to recognize, categorize, and put on a list. Hazards from electrical sources, for example, may be difficult for a novice to identify. However, high-power sources, transmission lines, and transformers provide a good basis for further assessment by qualified professionals in each particular technical area. Specialists in a wide variety of functional areas may be needed to accurately identify the levels of risk associated with different activities or systems.

Outline work processes involved in the business activity

Outlining the work process is a basic but necessary step to provide a broad perspective of the particular business or service activity. This is the time to step back, think, ask questions, and take notes. Questions should include the following:

- Which materials, equipment, machines, workstations, and work areas show injury or illness trends?
- Machinery and equipment have the potential to cause injuries and can be hazardous. Are hazard control devices built into the basic design of the machinery, or have machinery safeguards been added to protect workers who use the equipment? Is additional personal protective equipment—such as eye protection, gloves, or hearing protection—required to protect workers who use the equipment?
- Are chemicals used at the facility? Review the chemical inventory of the facility, as required by OSHA's Hazard Communication standard. Do any of the workplace injuries or illnesses come from areas where chemicals are used or as the result of chemical use? Safety Data Sheets—or SDSs—can provide clues to the problem and suggestions for correction or elimination of the hazard. OSHA's Hazard Communication standard mandates that chemical manu-

facturers provide SDSs on chemicals they sell to workplaces. Employers are required to maintain the sheets and keep them in an easily accessible place so that, in the event of an incident, SDSs are readily available to assist responders.

Organize the notes and retain these records as a guide to future activities. The purpose of carefully evaluating hazard potential from a broad perspective is to establish a formal management process that eliminates workplace hazards to the greatest degree possible or, when elimination is not feasible or possible, that recognizes that the hazard exists and sets up controls—such as safeguarding, employee training, or providing personal protective equipment—to provide workers with a large measure of protection from the hazard.

Focus on hazardous work areas

This step involves a methodical tour of the facility to plot work areas where the hazards occur. This activity is sometimes called *risk mapping*.

The following are examples of questions that can help focus risk mapping:

- In the chemical-mixing department, are visible dust, fumes, mist, or vapors in the air or accumulating on surfaces? Does ventilation capture all visible material and exhaust it from the workplace?
- In the packing-materials storage area, do potential fire hazards exist due to poor housekeeping?
- Are walking aisles clear of slip-and-trip hazards?

Input from workers, engineering managers, maintenance supervisors, and operations staff can add focus to these efforts. In addition, the loss prevention specialist or safety engineer of the workers' compensation carrier is an excellent source of information.

Some organizations have taken hazard detection to the next level by using a technique called *job safety analysis* or *job hazard analysis*. This process involves observing the job for several hours, listing the sequential steps required to perform the job, and identifying potential missteps.

Coming next ...

The next chapter will show how the probability and severity of hazards are evaluated and prioritized to define and describe various levels of risk.

4

HAZARD EVALUATION AND RISK ASSESSMENT

Once the operational process has been mapped and various hazards and exposures have been identified for the different steps, tasks, and work areas, it is possible to begin to prioritize which of the hazards need to be evaluated further. Because resources and time for evaluation are often limited, the safety professional needs to determine which ones should be quantified and addressed first. Armed with information about such criteria as the materials and quantities used, chemical toxicity, and duration and manner of exposure, a qualitative judgment may be made about the measurements and locations that should be evaluated and quantified.

Assess the degree of risk

Determining the potential exposure and hazards associated with the business activity is important because hazards do not present equal potentials for injury or damage. Some risks are more significant than others, and the frequency of occurrence of an incident or its adverse outcomes do not have equal probability.

Emphasis should first be placed on reducing hazards in areas where the worst losses have occurred and/or where there is potential for significant injury.

But beware: Assessing risk and assigning priorities can be subjective. Different people evaluating workplace injury frequency and severity records can reach different conclusions about hazard severity and prevention priorities.

For example, an operations manager may not consider a potentially hazardous machine a priority because no injuries have occurred to date. Conversely, a human resources manager might consider that same machine a big risk because staff have indicated fear of using it.

Risk is the product of the probability of an event occurring and the severity of the consequences of the event:

$$\text{Risk} = \text{Probability} \times \text{Severity}$$

By using the classifications from the tables below, it is possible to quantitatively identify for further analysis the jobs or tasks with the greatest risks. Based on company or industry statistics, job process observations, and SDSs for chemicals, it is possible to estimate the numerical ranking of each job area.

After reviewing the given work area and task, compare it to the information in Table 4.1 to evaluate where its probability falls according to its criteria descriptor for probability. Does the task take place infrequently throughout the day or not at all? Is the employees' work location so far away from the critical job process that you would consider them to have no potential for exposure? Find the associated probability and the numeric ranking for the job.

Table 4.1. Probability Ranking Criteria

Probability	*Numeric Ranking*	*Criteria Discriptor*
Frequent	5	This worker is subjected to this exposure or event throughout the duration of an eight-hour working day.
Likely	4	Exposures have been shown to occur on a regular basis.
Occasional	3	An exposure is likely to occur once in a while.
Seldom	2	Work practices and systems indicate little opportunity for employee exposure.
Unlikely	1	This worker in this task has no potential to experience this event or have this exposure.

Next, use Table 4.2 to evaluate the severity of the accidents that might occur at this job site during the intended work task. Would the scenario where this system broke down cause a catastrophic injury or death? Or might a failure cause possible short-term injuries and require first aid? Select the injury and illness severity for each job task along with a numerical ranking for severity.

Once the probability and severity rankings are estimated, the Risk Assessment Matrix shown in Table 4.3 can be used to identify where the particular job task or workstation falls in terms of consequence severity and the need for further assessment and remedial action. The jobs that fall in the chart near the top left should be evaluated further. Exposures should be quantified as much as possible and compared to Occupational Exposure Limits (OELs). These OELs may be in the form

Table 4.2. Severity Ranking Guide

Injury and Illness Severity	*Numerical Ranking*	*Criteria Descriptor*
Catastrophic	5	Death, loss of mental capacity, inability to work or function in society
Critical	4	Employee cannot return to workforce, may need to be rehabilitated for a new career; worker on extended leave and rehabilitation; loss of more than one workday
Marginal	3	Loss of one workday; worker receives medical treatment off-site
Negligible	2	Employee treated for minor injury at work
Insignificant	1	No consequence to employee or enterprise

of Permissible Exposure Limits (PELs) set by OSHA, Threshold Limit Values (TLVs) set by the American Conference of Governmental Industrial Hygienists (ACGIH), Recommended Exposure Limits (RELs) set by the National Institute for Occupational Safety and Health (NIOSH), or some other governmental or national or international consensus standard body.

Table 4.3. Risk Assessment Matrix

	Severity of Injury or Illness Consequence and Remedial Action			
Likelihood of **OCCURRENCE** or **EXPOSURE** for selected unit of time or activity	**CATASTROPHIC** Death or permanent total disability	**CRITICAL** Disability in excess of three months	**MARGINAL** Minor injury, lost workday accident	**NEGLIGIBLE** First aid or minor medical treatment
FREQUENT Likely to occur repeatedly	**HIGH** Operation not permissible	**HIGH** Operation not permissable	**SERIOUS** High priority Remedial action	**MEDIUM** Take remedial action at appropriate time
PROBABLE Likely to occur sometime	**HIGH** Operation not permissible	**HIGH** Operation not permissible	**SERIOUS** High priority remedial action	**MEDIUM** Take remedial action at appropriate time
OCCASIONAL Likely to occur sometime	**HIGH** Operation not permissible	**SERIOUS** High priority remedial action	**MEDIUM** Take remedial action at appropriate time	**LOW** Risk acceptable: remedial action discretionary
REMOTE Not likely to occur	**SERIOUS** High priority remedial action	**MEDIUM** Take remedial action at appropriate time	**MEDIUM** Take remedial action at appropriate time	**LOW** Risk acceptable: remedial action discretionary
IMPROBABLE Very unlikely; may assume exposure will not happen	**MEDIUM** Take remedial action at appropriate time	**LOW** Risk acceptable: remedial action discretionary	**LOW** Risk acceptable: remedial action discretionary	**LOW** Risk acceptable: remedial action discretionary

(AIHA, *Business Value Strategy Manual*)

Figure 4.1. An Industrial Hygienist Performing Workplace Air Monitoring

(Courtesy of Thomas P. Fuller)

Determining the potential exposure and hazards associated with the business activity is important because hazards do not present equal potential for injury or damage. Some risks are more significant than others; the frequency of occurrence of an incident or of the adverse outcomes does not have equal probability.

In certain cases of occupational exposure to hazardous agents such as noise, chemicals, or radiation, it is possible to evaluate exposure quantitatively. Measurements can be made by competent professionals, and such characteristics as chemical concentrations or noise levels can be compared directly with OELs. These measurements can then be used to determine the most effective and necessary systems or devices to be used to control the exposures and reduce the risks. Figure 4.1 shows an industrial hygienist performing an evaluation of air contaminants in a work setting.

Coming next ...

The next chapter describes the many methods used to eliminate or control workplace hazards.

5

CONTROLLING THE HAZARD

Wherever possible, the first steps taken to control risk should be the elimination of the hazard. This may involve eliminating a hazardous or toxic chemical from the process or substituting a safer chemical in place of the toxic one. Sometimes the process can be altered to reduce worker exposures to a hazard.

While eliminating hazards would seem the best answer in any situation, elimination isn't always possible. In these instances, controlling the hazard is required. As stated earlier, many methods to control hazards exist, depending on the work situation.

When controlling workplace hazards, there is a hierarchy that is typically followed in order to provide the maximum level of security to workers. The hierarchy begins with engineering controls that provide physical barriers between the worker and the hazard or some other system or equipment to reduce exposures. The second method of control is administrative. This includes the development of programs and procedures to protect employees. For example, training and warning lights are administrative controls. The last means of protecting workers in the hierarchy is the use of personal protective equipment (PPE) such as gloves, hard hats, and respiratory protection.

The following are common methods to control hazards, listed in order of precedence and effectiveness:

Design out the hazard

Review potential hazards in basic machine design, and design out the hazards wherever possible. "Design" in this context is used in its broadest connotation:

for equipment, facilities, hardware, tooling, materials selection, operations layout, and configuration.

While such hazard control is the responsibility of the machinery designer, purchasers of the machine also can demand safe design. Safe design will logically lead to safer methods, with fewer or no hazards if hazard recognition and control are inherent in the basic design process. For example, a hazardous saw built and designed with a guard to prevent employees from cutting themselves is much more effective than a guard that is attached retroactively.

Eliminate the hazard by substitution or automation

Some hazards can be eliminated by substituting nonhazardous materials for hazardous ones. Further, some hazards can be eliminated by automating a process with a high risk of hazard—for example, by isolating or separating the worker from the hazard wherever possible.

Engineering controls

Engineering controls use techniques and systems to reduce employee exposure to hazards. Such techniques include machine guarding and ventilation. This step reduces the hazard to a more acceptable level of risk and provides a safer work environment. Input of employees working these jobs can provide good ideas to reduce exposures.

Engineering controls include a system of devices and equipment specifically designed to provide direct protection of workers from potential hazards. Engineering controls are the first line of defense for workers since these controls are commonly permanent, secure structural systems that cannot be removed or degraded. They thus provide continuous maximum protection to ensure that employees cannot come in harm's way, regardless of their actions or system failures.

Some samples of engineering controls commonly used to protect workers include:

- Specially designed ventilation systems (e.g., general exhaust/fume hoods)
- Machine guards
- Barriers, shields, and enclosures
- Interlocks and automatic shutdown devices

Ventilation systems

Ventilation systems involve different types of fans and ducts that draw toxic fumes and vapors away from employees' breathing zones. General ventilation provides air circulation in a large area as a means of reducing the hazardous concentrations or providing clean air. Local ventilation control uses localized systems that draw noxious air away from the work zone. An example of local ventilation would be a fume hood in a chemical laboratory or a canopy hood over an open vat of solvents. A simple schematic diagram of a local exhaust system is shown in Figure 5.1.

Ventilation controls can be a very efficient means to reduce worker exposures to hazardous airborne chemicals and materials. These systems need to be designed and installed by knowledgeable, competent professionals trained in the field to ensure that the systems will work properly. All ventilation systems also must be tested and maintained on a regular schedule to ensure their continued effectiveness.

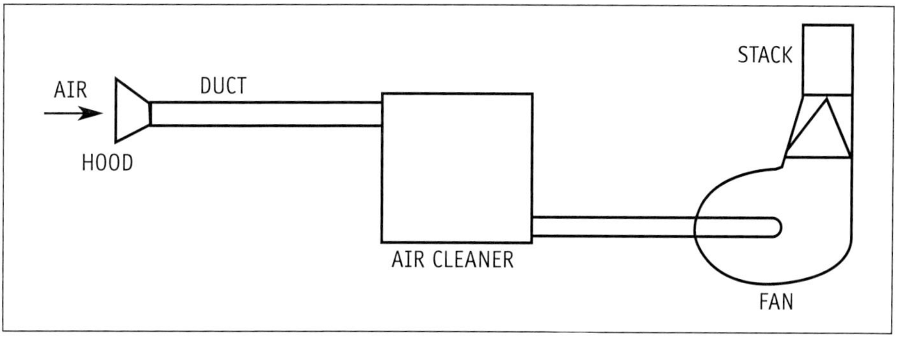

Figure 5.1. Schematic of a Local Exhaust Ventilation System

Machine guards

Equipment or machinery guards should already have been installed on new equipment by the original equipment manufacturer. Such protective devices include rails and barriers in construction, protection around the edge of platforms and roofing surfaces, covers over rotating parts of saws and maintenance shop machinery, and machine safeguards on production equipment.

Effective safeguarding, like hazard detection and control, is not a one-shot fix but rather a continuing process. Before any guard is mounted on a machine, an evaluation of guarding requirements of that particular machine needs to be made. This evaluation includes looking at such features as point of operation, flywheels and pulleys, air pressure, and hydraulic pressure. The process also must include employee and supervisory training in the effective use of machine safeguards.

Machine safeguarding is a starting point. But education and training of supervisors, mechanics, and operators also is vital. Most important, there must be a clear message from top management that effective machine guarding, like hazard detection and control, is in the mainstream of the company's operating culture.

Machinery never should be operated without first making an assessment of guarding needs and then installing the type of safeguarding that will protect the operator or maintenance staff effectively yet allow an optimum level of productivity.

"What kind of guarding do you need to produce quality work?" should be a question that supervisors and engineers ask those who work with these machines. A wide variety of guards are readily available that allow the operator to meet production and quality standards.

Common safeguards stop the machine if a hand or other body part is inadvertently placed in the danger area. Other devices work by restraining or withdrawing the operator's hands from the danger area by requiring the operator to use both hands on machine controls or by providing a barrier synchronized with the operating cycle of the machine that prevents entry to the danger area during the hazardous part of the cycle. These mechanisms also include restraining devices that pull the operator's hands back before the press cycles and presence-sensing light beams that shut down a press when a person approaches it too closely. Such devices are tamper resistant, sturdy, and allow for safe lubrication of the machine. In addition, they do not themselves create additional hazards.

Machinery manufacturers routinely supply guards on flywheels and other mechanisms on new equipment, but the problem for the equipment maker is that the ultimate use of the machine may not be known. Therefore, further safeguarding of the machinery is often the responsibility of the end user, who should have engineering or safety experts check out the machine to be sure that it has a safe operating process. The original equipment manufacturer can provide guarding for most applications, which is less costly than trying to retrofit the machine later.

Before purchasing new equipment, find out what standard guarding is available for the particular machine, and insist that the purchased machine be in compliance with OSHA's mandatory machine guarding requirements. By doing so, hazards can be eliminated or controlled by all feasible means, and residual hazards can be specified and warning labels affixed. The new equipment should include operator and maintenance manuals that contain instructions for safe operation and maintenance of the equipment.

A purchasing manager who understands the basics of safety engineering and is aware of compliance regulations should be alert to ensure that requirements are incorporated in purchase orders for new or remodeled equipment.

Barriers, shields, and enclosures

Often, a simple approach to worker protection is using a physical control to isolate workers from the hazard. A noisy machine can be isolated by enclosing it in a box to reduce the noise exposure to employees in the area. A protective shield could be installed under a conveyor belt carrying recycled metal parts to protect employees walking below. An operator of a crane used in a foundry can be enclosed in a small cabin to protect him or her from the excessive heat and noise associated with the process.

Interlocks and start-up/shutdown devices

Most machinery comes with traditional Start and Stop buttons or switches that control the energy going to the particular piece of equipment. Experienced operators know where these switches are and how to use them. They also know the dangers of trying to adjust the equipment when the power is still on. Life-threatening hazards come about when machinery power is not deactivated properly. OSHA estimates that some 3 million U.S. workers are at risk from uncontrolled hazardous energy.

A basic sense of how equipment works can be a building block to protecting employees who work with, operate, adjust, or maintain equipment activated by electricity or other forms of power. Employers should identify equipment-related hazards such as nip points, rotating parts, and flying chips and sparks. These hazards, when uncontrolled, have been responsible for many serious worker injuries and even fatalities.

Administrative controls

Written programs, policies, and procedures go a long way toward developing a safety program. Training employees about hazards and how to protect themselves from them is a very important part of any successful safety program.

The second line of defense to protect workers is administrative controls. These include written policies, programs, and procedures that are used to ensure systems are in place at all phases of the process to protect workers from hazards. Many administrative controls are required by OSHA, such as the Hazard Communication Program, Lockout/Tagout, Emergency Action Plans, Respiratory Protection Programs, and Hearing Conservation Programs.

Administrative controls assign responsibility for evaluation and control of all workplace hazards in a thorough and comprehensive manner. These documents become the basis for the entire safety program and are integral to its success and also provide the basis for hazard evaluation and documentation of exposure levels. They describe the safety systems in place to protect employees from specific hazards, and they identify the training requirements for all workers exposed to those hazards. Without written, comprehensive, and understandable programs and procedures, the other features of an effective safety program might as well not even exist.

Examples of various administrative controls include the following:

- Policies
- Programs
- Procedures
- Training
- Record keeping
- Warning signs and labels

Administrative controls are a series of programs, policies, procedures, and other nontechnical methods of reducing employee exposures to hazards. They are the second tier of protections because they tend to depend on the actions of humans to be effective. But they provide planned ways to conduct operations and ensure that processes are conducted and materials are handled in the safest ways possible.

Lockout/tagout procedures

One example of administrative controls is the OSHA lockout/tagout program. OSHA has developed procedures for safe shutdown, servicing, and maintenance of equipment in which an unexpected start-up or release of stored energy could cause injury to employees. In general, the regulations require employees to ensure that equipment energy sources are turned off or disconnected and that the switch or control is locked out with an employee-identified lock and/or a warning tag. The preferred and recommended method is the use of a lock.

Under the lockout/tagout program, employers are required to develop a hazardous energy control procedure meeting the following minimal standards:

- Develop a written energy control program.
- Ensure that new or overhauled equipment can be locked out during servicing.
- Develop and implement specific procedures including preparation for shutdown, the actual shutdown, equipment isolation, release of any backed-up or stored energy, and checking or verification that the isolation took place.
- Provide standardized locks and tags that indicate the identity of the worker using the equipment.
- Develop procedures for restoring equipment to service, including the release of lockout, notification of co-workers, safe positioning of employees, removal of the lockout devices, and clearance of materials and tools before start-up.
- Determine procedures for multiple persons servicing equipment and for work-shift or personnel change.
- Use additional methods to ensure safe work activities when tags are used instead of locks.
- Conduct inspections of energy control procedures at least annually.
- Train employees in the specific energy control procedures, with training reminders as part of the annual control inspections.

Increasingly, software for modern equipment has been developed that provides higher levels of control in the lockout/tagout program because lockout/tagout software can be integrated into facility or process systems.

Good housekeeping administrative controls

Another example of administrative controls is the use of good housekeeping programs. Most workers can recite the advantages of good housekeeping on the job—a cleaner, safer workplace; fewer trips, falls, and other injuries; better morale and productivity; and the facility being "a nice place to work." A chaotic workplace results in dirt and disorder, poor-quality work, misplaced tools and equipment, and slipping and tripping hazards.

A level of concern and interest about a neat and clean workplace is an informal building block of a broader program of housekeeping, which is an important ingredient in injury prevention. Neatness, orderliness, and good housekeeping in the workplace are basic deterrents to work injuries. Fire safety also is a natural result of good housekeeping, and an impending inspection by the local fire department can be an opportunity for good housekeeping training.

Custodians, cleaners, and janitors sweep and clean aisles and passageways, wipe away dust, remove all kinds of waste and trash, wipe off grease and oil from equipment, and empty trash containers. But the tidiness of an individual workstation, cubicle, or desk surface is often the responsibility of individual employees.

Many state or local building and sanitation codes set minimum standards of sanitation for washrooms and restrooms, legislate the minimum number of toilets, and spell out other requirements for washing facilities and restrooms. OSHA's Sanitation standard sets minimum-care standards for toilets, washing facilities, lavatories and showers, change rooms, and clothes drying facilities. Human resources managers are usually well versed in these laws. The best morale booster is a workforce that understands and practices individual responsibility for workstation tidiness. The custodial staff may be assigned to handle the rest.

The following suggestions can improve the effectiveness of housekeeping activities.

- Establish housekeeping responsibilities. A clear determination needs to be made at the management level regarding specific responsibilities for housekeeping and facility cleanliness. For example, individual employees are responsible for their immediate work areas; departments have another set of responsibilities; the custodial staff may have the responsibility for the main aisles and walkways, stairwells, building exteriors, and general plant equipment such as compressors and electrical rooms.
- Advise the custodial staff members of their specific housekeeping responsibilities, why the assignments are necessary, and how they can be carried out safely. Many facility managers provide checklists of duties for custodians so that the facility is cleaned on a regular schedule. Safety Data Sheets should be reviewed with custodial workers so that they pay attention to hazards encountered daily in their jobs. The review should include changing the technical words on the Safety Data Sheets into cautionary steps that employees should practice daily.
- Provide heavy-duty work gloves for janitors so that they can avoid cuts and abrasions. As basic as it sounds, require custodial staff to wash their hands thoroughly with soap and water before leaving an area that has hazardous materials and before and after using or cleaning the toilet, eating, drinking, smoking, putting on makeup, and handling contact lenses.
- Require that flammable liquids be properly stored in approved containers and metal cupboards for each type of hazardous substance.
- Provide covered metal waste cans for oily and paint-soaked waste or rags.

- Provide sufficient waste or trash containers for departments to collect trash and to ease the workload on custodial staff.
- Train supervisors to take the lead in housekeeping by keeping their offices neat and setting a good example by picking up litter as they encounter it.
- Food, beverages, smoking materials, drink containers, and food containers should be kept out of hazardous materials areas. Some organizations insist that all eating activities be conducted in lunchroom or rest areas.
- Avoid contact with contaminated clothing by keeping street clothing separate from work clothing. Place contaminated clothing in assigned containers.
- Make sure there is clear access to ladders, stairways, electrical switches, cutoff valves, fire extinguishers, and fire-fighting equipment. This requires regular reminders and follow-up.
- Keep floors clean and clear of debris. Wipe or clean up oil or grease from floor surfaces. Take prompt corrective action to eliminate any leaks at the source.

Personal protective equipment

Part of the hazard assessment process should involve a review of the need for personal protective equipment. When necessary equipment is identified, the employer should develop programs and policies for its use and require workers to wear it.

Personal protective equipment (PPE) does not eliminate the hazard; it only protects employees partially or fully from the hazard as long as it is used and worn properly. This equipment should be provided to employees working in areas where hazard exposure exists. In addition, workers should be trained in the effective use of the equipment and required to use it on the job.

Personal protective equipment is the last tier in the hierarchy of hazard control. This type of worker protection is the least effective means of protecting workers from hazardous exposures for a variety of reasons. When you think about it, much of the PPE used to protect employees is quite meager in comparison to the hazards to which the employees are exposed. A pair of gloves one millimeter thick is little assurance that a toxic chemical won't find a hole or seam by which it can pass through the glove and affect the worker. Similarly, the cartridge of a respirator is a thin barrier of protection for a worker potentially inhaling a hazardous gas or infectious agent.

Another reason why PPE is the last defense against hazardous exposures is that it depends on continuous and active use by the worker to be effective. Workers need to know how to wear the PPE for it to provide maximum protection, and they need to be diligent in their use of the equipment, which is often NOT the case.

There is often a misunderstanding that if PPE is worn, the hazards are eliminated. This is seldom true. PPE should never be provided or assigned without the oversight of a qualified safety professional. The level of protection of a respirator, for example, needs to be closely assessed in conjunction with the concentration of the hazardous agent in the air and adjusted to ensure that the OELs are capable of being met. Similarly, other PPE such as hearing protection and gloves must be assessed independently to determine the appropriate use and capabilities for each specific working condition. Some typical PPE is shown in Figure 5.2.

Figure 5.2. Personal Protective Equipment: Hard Hat, Eyewear, Respirator, and Earmuffs *(strods/iStock/ThinkStock)*

Personal protective equipment should be part of a comprehensive safety program that has evaluated all potential exposures. Only qualified, competent safety professionals should evaluate and assign PPE for any working condition and should do so only after all realistic engineering controls and administrative controls have been put in place.

Personal protective equipment includes all gear, accessories, and clothing that create a barrier between the employee and the hazard. This equipment can effectively protect the person only if it is of good quality, selected for a specific use, carefully fitted, worn where and when required on the job, and maintained in good condition *and* workers are trained in its use.

Despite its role, personal protective equipment does not eliminate a hazard but merely protects against it. Very simply, if personal protective equipment is not worn correctly—or if it fails—hazard exposure will occur.

To ease the introduction of new personal protective equipment items, supervisors and workers should be involved in the discussions so that workers buy into its use at an early stage. Employees are more likely to use items that they help select and that allow them to perform their work easily and with minimum encumbrance. A helpful purchasing department representative can often assist in selecting from the broad variety of available equipment.

Before buying personal protective equipment, the following questions should be answered:

- **Do hazards present in the workplace require employees to wear personal protective equipment?** OSHA regulations require employers to perform a written hazard assessment to determine the presence of hazards requiring personal protective equipment. If such hazards exist, employers are required to determine the correct type of personal protective equipment needed to protect their employees. OSHA also requires employers to provide protective equipment used regularly by employees on the job.
- **Is the exposure constant or occasional?** Machine operators in a factory may have constant, intense noise exposure during eight hours, whereas an office clerk may be exposed only for a short period of time while walking through the manufacturing area. The difference in exposure can influence the type of hearing protection needed. A supervisor who taps into a computer terminal several times a day does not face the same eye fatigue as a word processing operator who sits at a terminal all day.
- **How much protection is required?** A printer may need only gloves while mixing inks, but a chemical-mixing machine operator may require hand and arm protection, a face shield, and a chemical-resistant work apron on the job.
- **Does the job involve frequent bending or stretching?** The type of personal protection equipment needs to be assessed in terms of the work assignment.
- **Is the employee working in unusual environmental conditions?** Exposure to extreme heat, cold, or slippery surfaces each require different types of equipment.
- **What tools or equipment must the worker also wear?** A mechanic's toolbelt, for example, should be considered when assessing personal protective equipment needed.
- **Is double or triple personal protective equipment required?** Many workers need multiple types of protection.
- **How much quality do you want in the particular personal protective equipment item?** Items that are low cost but wear out quickly may not be a good investment.

Many employees don't like wearing personal protective equipment, finding it cumbersome. When employees understand why personal protective equipment is required, they become more understanding of the necessity of wearing it.

A short training program should highlight the reasons why protective equipment is necessary and when, where, and how it should be worn. Training should be documented. In some cases, such as with respirators and hearing protection, assistance should be given in fitting, cleaning, and testing the equipment. Equipment should be inspected regularly and defective items removed from service. Training can be reinforced when managers, supervisors, visitors, and contractor personnel wear their own personal protective equipment.

Personal protective equipment can be expensive. Generally, OSHA requires the employer to provide and pay for personal protective equipment required by the company for the workers to perform their jobs safely. Such equipment would include welding gloves, wire mesh gloves, respirators, hard hats, specialty glasses and goggles designed for working with lasers or ultraviolet radiation, face shields, and rubber gloves.

However, OSHA does not require employers to pay for equipment that is very personal in nature or that is often used away from the worksite. Examples of such equipment include nonspecialty safety glasses, safety shoes, and cold-weather outerwear of the type worn by construction workers.

Note that assessing the need for and providing personal protective equipment are only the first steps. Employers also are required to monitor and enforce the wearing of personal protective equipment, using disciplinary measures if necessary.

Determine priorities

The final step is to prioritize hazards based on the existence of imminent danger, historical losses, employee complaints, or employee inefficiencies. The first hazardous exposures to receive attention should be those with the highest potential for serious injury or even death.

Although the probability of occurrence may be low, highly hazardous exposures must be investigated and controlled promptly.

After the more serious exposures have been addressed, the remaining exposures should be prioritized, and resources should be allocated in descending priority based on the risk involved.

Coming next ...

The next chapter shows how supervisors play a significant leadership role in the success of a safety and health process. As representatives of management, supervisors communicate management policies, follow up on safe work practices, and, if necessary, enforce discipline for unsafe work practices.

6

SUPERVISORY LEADERSHIP IN SAFETY AND HEALTH

According to safety and health experts, attitude is critical in motivating employees to work safely. Attitude about safety—negative or positive—is reflected in the words and actions of top management and first-level supervisors alike. It isn't enough just to state the safety rules and processes; management also must follow and practice them. In short, management must create a corporate safety culture.

Supervisors are the direct representatives of management, and, therefore, employees interpret management's attitude as the supervisory attitude toward the safety program. To perform their jobs effectively, supervisors need training to understand their safety and health job responsibilities.

Helping the newly appointed supervisor

Newly appointed supervisors need to understand that hazard recognition and control and safe work practices are inherent parts of their new responsibilities. Many employees regard first-line supervisors as management because of the day-to-day direction supervisors give to employees.

Supervisors should be expected to perform the following safety and health duties:

- Work safely, providing the correct example to workers.
- Wear designated personal protective equipment on the job at all times.
- Give leadership to and support the facility's safety process.
- Correct unsafe acts observed or reported to them.
- Correct unsafe conditions promptly or arrange for prompt correction.

- Report unsafe conditions beyond the control of the supervisor to higher levels of management for prioritization and correction.
- Orient and train new or transferred employees in safe working procedures.
- Maintain a level of housekeeping in the department that minimizes incident potential.
- Arrange for periodic safety audits or inspections of the department.
- Investigate all recordable injuries and apply corrective measures.
- Arrange for injured workers to receive prompt first aid or medical attention.
- Educate workers on facility and department safety rules and insist that the rules be followed.
- Follow up with employees who are not working safely; retrain or redirect them in safe working procedures and gain their agreement to work safely.
- Hold employees accountable for safe work performance.

When unsafe work practices are observed, follow-up should be immediate, with the employee being told specifically what he or she did wrong. The employee should then be shown the correct work practice and asked to demonstrate it in the supervisor's presence. Additional follow-up should take place if needed.

Supervisory communication role

Good supervisors are good communicators, and once a safety and health process is in place, those safety and health activities must be communicated. Effective communication involves developing an easily understood message, delivering that message, and following up to ensure the message is understood.

Although the supervisor is the key person in the face-to-face communication of safety information, others in the organization—the human resources department, the graphics department, or the public relations office—can assist in nonpersonal communication.

Examples of communication activities can include:

- Reporting safety progress and concerns at the annual state-of-the-business meeting with employees and outlining the targeted safety goals for the year
- Using department staff meetings or start-of-shift toolbox meetings for safety reminders
- Using the safety committee to increase employee understanding of correct safety practices, employee involvement, and a sense of personal awareness about safety
- Including safety orientation in human resources orientation for new hires

Written communications can include the following:

- Handing out graphics, bar charts, and pie charts during a safety talk
- Using the company magazine or newsletter to explain safety activities and safety performance and to mention the names of those involved in safety committees
- Using safety suggestion systems to communicate safety concerns to management
- Posting safety banners at the plant gate and safety posters on bulletin boards
- Using a safety slogan to catch employees' interest and keep safety foremost in their minds
- Sending notices of safety achievements to local newspapers

Communicating the message

Managers and supervisors, especially first-line supervisors, have the basic job responsibility of communicating with the workforce. The following tips can help you get across your message about safe job performance:

- Keep employees informed on matters of safety, quality, production, and cost that affect their jobs, the department, and the business. This will bolster understanding, motivation, and job performance.
- Speak slowly and deliberately to generate an attentive but relaxed atmosphere that invites questions. In the middle of the presentation, restate slowly what is happening and sum up at the end.
- In some cases, to limit the chance of misunderstanding, it may be prudent to hand out a summary that follows very closely what is being said. When a large number of workers are affected, repeat the message through an e-mail system, or offer a toll-free telephone number inviting employee questions.

New hire orientation and safety training

Safety and health experts emphasize that more than 80 percent of workplace injuries are caused by unsafe acts and only 20 percent by unsafe work conditions. The risk of injury to newly hired employees is much greater than that for experienced employees.

One-on-one safety training by a job instructor, mentor, or supervisor is one of the most effective methods of job instruction because it catches a new employee when that person's interest and motivation to learn the new job are at their highest.

Handling employee safety complaints effectively

Managing employees entails listening to employee questions, concerns, and complaints; investigating them; and doing something about legitimate safety complaints. This is important because unsatisfied or unaddressed employee safety concerns can result in injuries or in the complaint moving to higher levels of appeal inside or outside the organization.

Management should develop a procedure for receiving and handling employee safety and health complaints. Publicize the procedure and make sure it works. Employees will respect an active safety and health process that involves them, corrects hazards, and reduces injuries.

Encouraging use of an internal complaint process

Management should encourage employees to use internal complaint procedures to make their concerns known. The following suggestions maximize the effectiveness of an internal complaint procedure:

- Workers should be encouraged to come forward with their workplace safety concerns, complaints, and suggestions.
- Supervisors are usually the first to be aware of a safety and health problem and should be trained in the company's complaint procedures.
- Supervisors or their managers should provide communication to employees about what was done to investigate or correct a situation. Employees should be advised of the time period required to correct a condition.

Complaining versus insubordination

A fine line distinguishes legitimate employee complaints and insubordination or refusal to perform assigned tasks. An employee who airs a complaint but carries out assigned work technically is not guilty of insubordination. The safety coordinator should seek guidance from the organization's human resources professionals regarding the handling of insubordination.

Generally, employees are required to "obey now, grieve later" when it comes to complaints about work assignments or other supervisory decisions. They cannot hide behind a complaint in order to dodge their work responsibilities, except when performing a work assignment would endanger their own or a co-worker's safety and health. Many state laws and court decisions support this approach.

The employee must be able to show that:

- A reasonable person would conclude that a real danger of death or serious injury existed
- Insufficient time existed to eliminate the danger through the use of regular OSHA procedures
- The employee attempted, unsuccessfully, to persuade the employer to correct the dangerous condition

But if an employee's outburst directly undermines a supervisor's authority, the behavior may rise to the level of insubordination. These cases are not easy to sort out; safety concerns may sometimes be convenient levers for employees or unions to pursue other workplace objectives. And, of course, individual employee opinions on what is safe or unsafe are highly subjective and may differ from those of supervisors, technical experts, and management.

The supervisor's role in enforcement of safety rules

Safety rules are designed to ensure order and encourage incident-free workplace behavior. Effective rules must be realistic, fair, and easily understood. Successful safety rules are those developed with supervisor, employee, and safety committee input. Some organizations even ask for input from all employees; doing so encourages employees to modify their work habits before the new rule is even finalized. Employee, supervisory, and union involvement in the development of safety rules results in the principal supervisory role being one of reminder and follow-up.

A progressive discipline structure should be adequate to handle safety violations, unless the infraction is so serious and injury potential so high that the employee should be suspended from work or terminated. When employees see that supervisors intend to enforce safety rules even to the point of discharge, compliance will readily follow.

Types of safety rules

Two types of safety rules generally are used: general safety rules for the guidance of all employees and safety rules specific to the department or job. *General safety rules* are those covering many aspects of safe working behavior in the facility and apply to all personnel. Such rules may deal with smoking/no smoking areas, walking/running, use of general personal protective equipment such as safety glasses and hearing protection, and machine guarding measures across many departments. *Specific safety rules* are those dealing with the safe work practices of a particular department, job, or work assignment.

Unsafe work practices

Supervisors play a vital role in fostering the type of work atmosphere that encourages safe work habits and detects and corrects hazards. But despite efforts to establish a safe workplace, some workers ignore these rules.

What is a supervisor to do when faced with employee misconduct? Management policies, supervisory meetings, and employee orientation and training should address the consequences of ignoring safety rules and policies. If an employer tolerates violations of safety rules or OSHA standards, then the employer is accepting responsibility for the results of those acts. Employers should make sure that rules are stated clearly and understood, that employees are well trained in proper procedures, and that employees understand that violations of safety rules will not be tolerated.

Discipline should be applied as uniformly as possible. The report of an incident is often reviewed with higher levels of management so that uniform discipline is applied for like infractions.

The employer is responsible for just about every work-related injury or illness, but exceptions do exist. For example, employee misconduct on the job may be the basis for denial of workers' compensation benefits under some state laws. A safety coordinator will need guidance from a human resources professional in handling these situations.

Dealing with the "incident repeater"

Injuries incurred by so-called incident-prone employees exceed those of other employees who do similar jobs. Their behavior includes frequent trips for first aid or medical treatment, violations of safety rules, off-the-job injuries that affect job performance, and unsafe work practices that regularly need correction.

A variety of personal health, psychological, and physiological reasons can contribute to the problem of one employee averaging more incidents than other employees. Although it is easy for supervisors to jump to conclusions about incident repeaters, it's better to try to identify the problem so that fact-based corrective action can be taken. Managers should consider the following factors:

- How high is high? The number of injuries can be measured through a review of the first aid log, the OSHA 300 Log, and workers' compensation reports. Is it the individual or the job that is causing the injuries? From these logs, department injury averages can be generated so that individual records can be compared to these averages.
- Is it an individual problem or a problem of a group of workers doing the same job? That should lead the manager to investigate the job and search for potential hazards and what can be done to correct them.
- Is the employee exposed to hazards over which he or she has little control? Employees don't design or install machine guards and usually have little say about potentially dangerous equipment before it is installed. The facility's hazard evaluation team needs to investigate, evaluate, eliminate, or control the hazard.
- If worker incentive plans are in effect, could a worker be trying to exceed productivity norms to earn a higher bonus?
- Are off-the-job pressures affecting the person's attention on the job, or is the individual's health deteriorating and affecting job performance?
- Is retraining necessary? Will temporary assignment of a job instructor, mentor, or trainer help correct the behavior?
- Is a transfer to a less-demanding job appropriate?

Some employers include incident repeaters in safety activities such as a safety committee. Such involvement sometimes generates increased interest and corrects the problem. But make sure that an incident repeater does not detract from the positive efforts of other safety committee members.

An allegation of being an incident repeater should be focused on an individual basis. A discussion should be held with the employee to review the information accumulated, emphasize company safety policies, and discuss how the unsatisfactory behavior can be changed. The worker should be asked, "What do you think needs to be done to correct the problem?" Results of the meeting should be documented.

Employers should be prepared to take action against those workers who sustain repeated preventable injuries because these incidents are legitimate work performance issues that should be addressed through the organization's corrective action process. Incident repeaters subject to disciplinary action clearly should have incident records in excess of the average.

With careful effort and attention, most employees who are considered incident repeaters can learn positive attitudes and adopt safe work behavior.

Adjusting supervisory styles to different employees

Sometimes the answer to correcting employee performance and safety problems lies in the supervisor adjusting his or her style of supervision to suit the particular employee.

The challenge is to get the supervisor to manage the employee in the safety process legally and fairly and with confidence. Techniques will vary with the particular employee involved—from the new hire during a break-in period on the job to the respected employee whose performance is declining.

Once a supervisor has found the reason or reasons for an employee's non-productive or unsafe behavior, he or she must take corrective action. This may involve counseling, reinstruction in job duties, redirection of work efforts, or a change of assignment or shift.

Most organizations have established procedures to handle errant employees when basic efforts at correction haven't worked. If counseling and cautions have not succeeded, more serious and formal procedures may have to be taken.

The good efforts of the majority always must take precedence over those few who are unable or unwilling to meet workplace objectives. Employees who respect safety rules and measures need to have their efforts rewarded with merit increases, production bonuses, and individual and group recognition. Those steps help keep that productive engine moving along the track of organizational success.

The supervisor's challenges are to keep calm under pressure and to generate high levels of morale among staff. The supervisor should act as a cheerleader who encourages good work and supports and encourages workers when setbacks are encountered. Morale remains high when employees share a positive sense of spirit and are willing to help an organization achieve its goals.

Coming next ...

The next chapter will show that valuable resources to assist incident prevention activities exist in every workplace. This resource is the organization's employees. Management has the ultimate responsibility to make and keep the workplace safe, but employee involvement in the safety and health process can help achieve that responsibility.

7

INVOLVING EMPLOYEES IN INJURY PREVENTION

Most people will not consciously take chances with their own safety. The same holds true for employees, particularly if they understand the reasons for the safety rules and are trained to follow these rules as part of their job duties.

This basic employee self-interest can be a prime motivator to engage in a safety activity. Further, the Occupational Safety and Health Act requires all employees to comply with OSHA standards and rules, to wear or use prescribed protective equipment at work, to report hazardous conditions to supervisors, and to report any job-related injury or illness and promptly seek medical treatment.

Whether prompted by self-preservation or the existence of a standard or regulation, employee commitment to safety can be increased with involvement in safety activities. For example, employee participation in safety and health committees and safety circles raises awareness of correct safety practices and highlights unsafe work procedures needing to be corrected. An organizational culture of safety that encourages continuous improvement will go a long way toward achieving an effective safety program.

Safety committees

Safety committees are an important way for employers to demonstrate to employees their interest in safety. In addition, rotating committee membership over time allows many employees to become involved in incident prevention activities. Employee involvement increases on-the-job safety consciousness and commitment to safety.

Safety committees will fail under the following conditions:

- Waning management support
- Incomplete projects on the agenda for long periods of time
- One or two members dominating meetings
- Underfunded committee activities
- Boring meetings bogged down in detail
- Declining attendance, overloading remaining members with assignments

An active safety committee also can reduce workers' compensation premiums. In fact, about half of the 50 states have established some type of requirement for the establishment of safety committees.

An effective safety committee can assist management with identifying and controlling hazards, defining corrective actions, and investigating incidents.

To assume a major role in hazard detection and control activities, safety committees need proper direction and support from management. Employers should think through the safety committee's structure and role. Such preparatory work will pay off in more effective results from the committee's efforts.

Consider the following questions:

- **Is management willing to share vital information, including workers' compensation costs and rates and incident frequency compared to similar companies?** This trust is necessary for effective employee involvement in the incident prevention process and necessary to build and lead a team.
- **Why is the safety committee being formed?** Focusing on the reasons for establishing the safety committee should lead logically to determining the precise role of the committee in assisting management in hazard control efforts.
- **Will the safety committee activity be part of a total safety process?** A safety committee structure acting as a stand-alone safety activity, and not integrated into the total safety process, will likely fail.
- **What are the committee's goals and objectives?** This may involve discussions with supervisors, union representatives, or employee leaders, but ultimately, management must agree on how the committee will operate.

Two important questions need to be considered and answered before the safety committee members are announced.

- **How will the safety committee be held accountable?** With a clear charter, training, and direction, management should expect productive efforts from the committee in terms of incident reduction, hazard detection, and control procedures.

> ***How to select safety committee members***
> Members of successful safety committees have exceptional qualities. They are motivated to work safely and to help others work safely. They are regarded by supervisors and co-workers as safe workers and leaders whose examples will be followed. They have knowledge of specific work operations in the department or division and are tactful and diplomatic communicators. They are dedicated and attend safety committee meetings regularly, participating actively in discussions. They know where to report unsafe conditions that can't be corrected on the spot. They are willing to work with others toward the common goal of a safer workplace. They understand that complaints and grievances not relating to safety do not belong in safety committee discussions.

- **How large should the committee be?** An effective safety committee should include up to seven people and include a good balance of employee and management representatives.

Getting started

The first safety committee meeting should include decisions about frequency of meetings and committee member responsibilities. Safety meetings should not be held on Fridays because any action items identified and agreed upon cannot be acted upon until the following week. Effective meetings are held earlier in the week and allow action to be taken in the days immediately following.

Further, the committee should determine how the results of safety committee activities will be communicated to the organization. For example, minutes of safety committee meetings should be distributed to management, supervisors, and safety committee members and posted on notice boards to help publicize the committee's positive role. Co-worker recognition will motivate members to continue their good work and will keep all employees informed of the progress of the company's incident prevention efforts. The good news about safety successes also becomes a recruiting tool for future safety committee members.

Once the committee is operational, a safety committee meeting agenda should be issued several days before each safety committee meeting. This will allow members to focus on the discussion planned for the meeting and is an excellent follow-up tool to get projects and reports completed by previously agreed-upon due dates. An agenda also serves as written documentation for top management by relaying the status of safety activities taking place.

Good meeting agendas also can help ensure that the meeting can cover necessary information in an efficient amount of time without overloading the participants. The committee's emphasis should be on the steps necessary to prevent workplace incidents or repetition of the same type of incident and on the positive initiatives taken to lower unsatisfactory safety statistics.

A typical safety committee agenda

1. Review the minutes of the previous safety committee meeting and report on completed items. Assign new due dates for incomplete items.
2. Review recordable injuries, lost workday cases, and near-miss incidents since the last meeting. Issue follow-up recommendations and actions to be taken.
3. Review incomplete items on safety reports. Review safety-oriented work orders in engineering and maintenance departments, the priorities assigned, and the dates set for completion.
4. Report safety successes—unsafe conditions noticed and eliminated or corrected and unsafe work practices stopped and corrected.
5. Discuss new safety business and report on special safety projects.
6. Present education subjects and determine topics for the next meeting.
7. Adjourn and announce the date and time of the next safety committee meeting.
8. Distribute minutes of the meeting within 48 hours.

Charting the committee's progress

Safety committee activities should be monitored for progress. The best-laid plans of management may not translate into an effective safety process. Safety committee members should be able to provide feedback to adjust safety efforts. Organizations with successful safety activities review the planned program at least semi-annually, sometimes quarterly, and make any needed corrections to the process.

Monitoring the committee's progress also will pinpoint areas where continuing education is needed. Safety committee members know their own jobs well but don't necessarily know effective safety management. Training and education can do a great deal to enhance committee members' knowledge, making them more interested, involved, and effective.

To increase committee participants' understanding of their roles, a subgroup of the main safety committee can be charged with developing an education topic for each meeting. The subgroup may use in-plant experts or outsiders—such as insurance company loss control representatives or vendor representatives—to educate members and increase the committee's effectiveness.

OSHA area offices also provide speakers on safety and health subjects for such meetings. These presentations should be short (maximum 20 minutes) and include audiovisual training materials and summary handouts that committee members can take back and use on the job.

Safety circles

Greater employee participation in many aspects of the business has been the hallmark of business operations in recent years. Such participation can be seen in corrective action teams and quality circles in which management seeks employee input on improving operations and reducing defects.

These same principles of employee involvement can apply to safety. Such employee groups are often called safety circles. The same principles outlined for planning the development of a safety committee can apply to safety circles.

Safety circles identify a problem, collect and analyze data, determine the most logical solution to the problem, and present to management the recommended solution and the costs (savings or expenditures) involved. Safety circles typically tackle workplace conditions such as causes of arm and upper torso strain, the effectiveness of machine guarding, and the coordination between maintenance service staff and department mechanics in implementing lockout procedures.

Gaining union involvement

Employer safety abuses in the 20th century provided many unions with their original appeal to workers. Today, some labor contracts spell out specifically what management and the union will do about safety and employee participation in safety activities. Employers can normally expect good support from a labor union regarding positive management activities to promote a safe workplace.

Some organizations with unionized workforces establish union-management safety committees or joint safety committee structures. These committees operate much like a typical safety committee, except that the union may have greater authority with respect to tasks such as designating employee committee representation and indicating the process for presenting and acting on safety improvement proposals.

While unions generally support joint safety and health committees, the National Labor Relations Act prohibits employers from dominating a group of employees and from dealing directly with employees about conditions of employment. For this reason, safety committee coordinators should obtain guidance from a human resources professional or employment law attorney to ensure that safety committee activities do not infringe upon the law.

Management should communicate company injury rates, safety problems, hazard detection and control efforts, and workers' compensation costs to unions. Sometimes there is insufficient understanding of the importance of a good safety program in reducing costs, injuries, and accidents. Breaking those costs down by plant or department can help focus the subject and demonstrate that the areas with the best safety programs actually have the lowest costs. Workers' compensation costs can be broken down to a cents-per-hour cost for each hour worked to show the high impact these costs have on total business costs.

Management's concerns about workers' compensation losses also need to be taken to the main union committee, including paid union staff representatives. Such communication helps pave the way for tighter safety and health rules.

Safety rules that work

Management is responsible for setting standards for product size and shape, quality parameters, and engineering specifications and tolerances. Management is also responsible for determining safe operating practices for its business or service.

Like any set of practices, safety rules are designed to ensure order and direct employees toward incident-free work behavior. But, as with any set of rules, workers can resent being policed and may comply only when supervisors are present.

To ensure success, safety rules should be developed with employee input, rather than written by people who are unfamiliar with the jobs or the associated hazards. Planning and discussion with workers and supervisors during the preparation of safety rules will lead to much greater acceptance of these common-sense rules. Usually, two types of safety rules exist:

1. General safety rules for all employees of the facility
2. Safety rules specific to the department or job

General safety rules cover many aspects of safe working practices in the facility and apply to all personnel. Such rules may deal with walking/running, use of personal protective equipment such as safety glasses and hearing protection, machine guarding, emergency evacuation procedures, and smoking/nonsmoking areas. Specific safety rules deal with the safe work practices of a particular department, job, or task.

Employee input is a critical element to formulating safe work rules that will be understood, accepted, and used. Worker involvement in devising safe work rules draws on their thorough knowledge of their own jobs and motivates them to adhere to those rules. Supervisors also should be asked for input because ultimately, they will be responsible for enforcing the rules in their departments.

Draft safety rules should be reviewed and approved by an experienced safety professional and, because of the possible employee relations implications, by a ranking labor relations or human resources professional. Rules should be reviewed with a union committee prior to publication because rules affect working conditions, and union representatives likely will be involved in any disciplinary proceedings that occur.

Communication and enforcement

Organizations should develop and maintain formal means for safety communication that enable a continuous safety improvement loop and ensure that all personnel are fully aware of the safety program.

Safety communication should aim to:

- Ensure that all staff are fully aware of the safety program
- Convey safety-critical information
- Explain why particular actions are taken
- Explain why safety procedures are introduced or changed
- Convey "nice-to-know" information

Safety communication could take various forms. The internal (to the organization) means of safety communication include, but are not limited to:

- Safety bulletins
- Safety notices
- Newsletters
- Briefings
- Seminars and workshops

- Refresher training
- Intranet

Safety communication also encompasses the distribution of safety procedures within the organization.

Examples of external means of safety communications are:

- Meetings, workshops, and networking
- Websites, online fora, blogs, and e-mail distribution lists
- Magazines and other publications

Safety communication is an important enabler for improved safety performance. Safety lesson dissemination is a vital element of safety communication because lessons learned from past experiences that are implemented within the organization reduce the chances of accident and incident recurrence. Many factors such as time, distance, environment, and a sense of complacency often result, unfortunately, in lessons being either ignored or forgotten. Lesson dissemination could be considered an indirect form of safety training.

Once finalized, safety rules should be communicated to employees. An effective way to initiate a new set of safety rules is to review them with small groups of employees at meetings and provide an opportunity for employees to ask questions. Employees affected by the rules should be given a copy. The rules also should be publicized through the workplace newsletter and posted on bulletin boards. Specific work rules for departments or jobs should be given to each worker and posted prominently in their work areas. A copy of the rules should be part of the company supervisor's policy manual as well.

In larger organizations with intranet resources, company policies and safety programs can be maintained on internal or public locations. Employees with computer access can obtain, read, and refer to documents almost instantaneously. As systems, programs, policies, and procedures are upgraded or updated, it is easy to change and update the website.

All new hires should be provided with a copy of the safety rules and be asked to acknowledge in writing that they have reviewed and understand the rules.

Safety rules are established for the safety of all, including managers and supervisors, outside contractors, and visitors to the facility. Clear communication should be issued stressing that the rules apply to all employees and nonemployees entering the facility.

If good employee, supervisory, and union involvement has taken place in the development of the safety rules, the supervisory role in enforcement of the rules primarily will be that of reminder and follow-up. When employees see that supervisors intend to enforce safety rules even to the point of discharge, compliance will follow.

Reprimand for infractions should follow as soon as possible after an incident, with the employee being told specifically what he or she did wrong and why it was wrong. The degree of discipline to be applied should be determined, the discipline explained, and the employee urged to correct the improper practice.

Management responsibility

Some organizations set up a safety committee as the safety process. But safety committees or other forms of worker participation alone rarely prevent injuries. A safety committee that is properly directed, trained, empowered, and provided with the financial and staffing resources to carry out assignments can assist management in meeting its safety responsibilities. But legally, management has the ultimate responsibility for safety. For this reason, an increasing number of safety-conscious employers are forming a committee in addition to the facility safety committee. The safety steering committee at the management level gives oversight and leadership to the company's safety efforts.

The key players in company safety activities manage the overall process; establish general policies, procedures, and inherent programs; and assign responsibilities, ensuring accountability for all safety activities. Action items from the safety committee that get bogged down or need management action require the attention of this decision-making management committee. With a sound safety process underway, management steering committees have been found to be highly successful in spearheading hazard control efforts.

Coming next ...

The next chapter will show how effective, informal safety activities can be brought together into a total safety and health process in which roles and responsibilities of workers, supervisors, and managers are delineated; priorities are determined; and time frames for hazard control and elimination are assigned.

8

FORMALIZING THE SAFETY AND HEALTH PROCESS

Some organizations start off their annual safety drives with a goal such as "Zero incidents this year!" As much as many managers and employees would like to publicly state a goal of zero incidents, such a pronouncement may not be practical. If a great deal of publicity surrounds such a slogan, a real letdown can occur after an injury.

As a result, some organizations stay away from the zero approach to safety targeting. An increasing number of forward-thinking organizations have moved away from a zero-incidents goal, preferring to promote hazard detection and control activities as the core of their safety and health processes.

Setting goals and objectives

The objective of safety planning is to achieve a continuous improvement of an organization's safety performance. In order to achieve this improvement, safety performance indicators must be developed to measure and demonstrate that the achieved level of safety meets the targets.

The principal safety objective is a qualitative or quantitative statement that defines the aspirations and strategic goals of an organization and relates to the safety of organizational activities, services, and products.

> Exceptional organizations set their objectives formally—clearly enunciating their vision, defining desired outcomes, spelling out the attainable steps for meeting the objectives, and documenting the process (ICAO Doc 9859—*Safety Management Manual*).

The acceptable level of safety can be expressed qualitatively in general terms. Or, more quantifiable goals may be used in the form of safety performance targets. Qualitative goals may be described in such terms as "maximum achievable safety." These are nonquantifiable measures that indicate a direction and overall philosophy for the development of a culture of safety in the organization, at all levels. Quantifiable goals identify the specific safety targets that are sought, such as reduced accident or injury rates. The number of regulatory violations or audit deficiencies are safety targets that can be measured, tracked, and trended. A safety performance target comprises one or more safety performance indicators as well as desired outcomes expressed in terms of those indicators.

Sometimes referred to as "goals" or "objectives," the safety performance targets are determined during the planning phase. They are established in order to ensure the achievement of the acceptable level of safety that is considered desirable and realistic for the individual operator/service provider. The desired safety outcome (target) may be presented either in absolute or in relative terms. An example of a desirable safety outcome communicated in absolute terms is "less than 1 serious injury per 1,000,000 operating hours."

A wider range of different safety performance targets (and indicators) will provide more useful information than the use of a single target. The relationship among an acceptable level of safety, safety performance targets, safety performance indicators, and safety requirements is as follows: An acceptable level of safety is the overarching concept; safety performance targets are the quantified objectives pertinent to the acceptable level of safety; safety performance indicators are the measures/metrics used to determine if the acceptable level of safety has been achieved.

Safety performance targets are subject to a periodic review and update, as necessary. These reviews are carried out as part of the strategic safety planning and improvement activities of the operator/service provider.

Safety performance indicators are used to measure the level of safety performance achieved in a system and are linked to the safety performance targets. They enable the organization to measure and demonstrate the achievement of the set target levels. The safety performance indicators should also be easy to measure and accessible over time.

In general, safety performance indicators are presented in terms of the frequency of occurrence of harmful event(s). Safety indicators differ among different industries. Some industries have readily accessible lists of performance indicators already developed. These lists can be used for comparison among different organizations.

Safety planning

Once safety program goals are identified and the targets have been established, programs and procedures need to be developed to address those targets. Writing out programs and procedures to address specific safety regulations is often a good place to start.

Regulations often outline a specific set of criteria to be met in terms of performance or systems. OSHA clearly describes what information needs to be in a Hazard Communication Plan. The organization's Hazard Communication Plan should be written to OSHA specifications to ensure the regulations are being met. Where there are variances from the regulations or standards, the rationale should be described in the documents.

In other aspects, OSHA may not give a specific formula for meeting the regulations but leave it up to the organization to determine how meeting the regulations is to be accomplished. For example: OSHA says injuries and illnesses need to be recorded and reported; it is up to the organization to develop the programs and specific procedures to describe how this is to be accomplished. That is, where do workers go to seek medical attention and report an injury? Who is responsible for keeping the records and sending the reports to OSHA? How is this information to be distributed within the organization? Unless the recording and reporting systems are described in detail and provided in writing, it is unlikely that they will be adhered to consistently.

Putting it in writing

Putting information in writing is a core business activity. Summarizing existing safety and health activities in a written format formalizes what is working well, allowing management to see in a single document the facility's injury prevention plans. The written plan also designates responsibilities and accountabilities, priorities, and the time frame to accomplish the objectives.

The document need not be long. It should list goals and objectives and should cover the essential elements of an effective safety process, how the facility intends to implement that plan, who is responsible (management, supervisors, safety committees, or employees) for key elements of the plan, and the time frame set up for completion.

A written safety action plan is a guide for the organization into the next period or time frame, typically one year. At year-end a written plan can become the foundation on which to build the program for the ensuing year or other time period, again amended or expanded to incorporate the lessons learned in the past year.

Putting the plan in writing forces planners to think through the key elements: What is going well and what needs improvement? Revising the written action plan each year requires planners to review hazard control experiences in the past year, modifying where needed, strengthening weak areas, and reinforcing elements that worked well. Once the basics of a safety and health process are in place, management will be able to set annual goals and objectives—the true benchmark and scorecard process for checking the success of safety efforts.

Although management should lead the development of a written safety plan, many other people (managers, supervisors, safety committees, employees, and outsiders) can assist. The plan should be a hands-on document detailing the proactive safety steps the facility is taking and specific plans for an ensuing time period. It should include hard facts, specific plans, and firm intentions.

In recent years, several business models that relate to safety have become prevalent. These models can be integrated into existing management systems or used as the basis for a new program. Underlying principles of each model include rigorous assessment of the safety program and the goal of continuous improvement in all aspects. Some of the more common management methods are listed briefly.

Safety Management Systems (SMSs)

Over the past several years, more sophisticated safety and health management systems (SMSs) have been developed to provide systematic approaches to managing safety. The programs developed provide a methodical approach that can be followed by any organization to achieve safety goals. An SMS involves setting goals, planning, communicating, implementing, evaluating, and continually improving. The SMS approach uses an organized framework to establish and promote the positive corporate safety culture.

The SMS approach begins with a safety policy. This should be a clear and concise statement about the company's philosophy and approach to safety, with an indication of the goals to be achieved. It is the first important milestone of SMS implementation and defines the value of safety in the overall business and performance framework of the organization. Ideally, the safety policy should corroborate with the organization's commitment to safety and clearly indicate that safety is afforded the highest priority in the service provision.

The safety policy may take many different forms, but normally it is a written document that describes the generic principles upon which the SMS is built and operated. A typical safety policy document would consist of a policy statement that is further expanded by a number of basic safety management principles to be followed, namely, commitment to safety, safety priority, safety responsibility, planning for safety, safety management, safety standards, safety achievement, safety assurance, and safety promotion.

The policy statement is a declaration of the overall safety objective of the organization. It sets the general direction and aspirations for improved safety performance.

The safety policy statement may build upon the following basic safety principles:

Commitment—The organization's senior management is fully committed to safety and will be fully supported in achieving program goals.

Responsibility—All staff and management of the organization have a shared responsibility for achieving safety goals.

Planning—Programs and procedures needed to achieve goals and targets must be developed, written, and integrated into organizational systems. Compliance with regulations and industry standard best practices are an integral part of all safety programs.

Evaluation—Resources and methods need to be developed and implemented to measure and evaluate the effectiveness of safety programs. Surveys, records, audits, and reviews should be used to identify deficiencies and weaknesses that can be corrected to achieve goals, objectives, and performance targets. Recommendations for improvement are identified, and corrective actions are developed and completed on schedule.

Promotion—Resources, time, and energy are made available to effectively communicate the safety program to all employees throughout the organization. Open communication is encouraged to promote safety goals and the safety culture.

Continuous improvement—As goals are achieved, new safety goals are identified. Programs and procedures are revised to reflect new objectives.

OSHA recommends—but does not require—that an employer's safety program be in writing, although some OSHA standards (including Hazard Communication, Confined Spaces Entry, Bloodborne Pathogens, and Process Safety) require a written implementation plan. During OSHA inspections, compliance officers will check for a written plan. A written acceptable safety document also will be considered when assessing fines against an employer and may influence the reduction of those fines. OSHA provides a program evaluation profile listing the basic elements its compliance officers look for when examining an employer's voluntary efforts to develop a safety and health plan (see *www.oshaslc.gov/SLTC/safetyhealth/pep.html*).

- Preparers of a written safety program should do the following:
- List the time period to which the program applies, typically a year.
- Specify the geographical coverage of the safety plan.
- Crystallize the intent of the process: What is your program trying to accomplish?

Examples include protecting employee safety and health, reducing the number of incidents, reducing workers' compensation costs, addressing OSHA concerns following recent inspections, and focusing on specific problems your hazard evaluation has uprooted.

- Set goals. In an initial plan, the basic structure for a hazard detection and prevention activity should be a principal goal. With experience and time, these goals may involve the overall reduction of injuries to a lower statistical level or a focus on one type of injury, such as cumulative trauma disorders or safe handling of chemicals.
- Determine who will write the plan. The most meaningful safety plans are those prepared with safety staff input, including those involved directly in the local incident prevention activity.
- Select the ways the plan will be communicated. Consider having copies made for supervisors and the safety committee that include highlights in the facility newsletter and regularly refer to the plan in safety committee meetings. At the start of each new plan year, supervisors should explain highlights of the updated plan in crew meetings.
- Stress accountability, including implementation or completion dates.
- Calculate costs. Most organizations insist that the costs of a formal safety process be documented. Others develop a budget for the number of work hours required in safety meetings or the number of hours away from the job for safety inspection tours. The safety budget also may include the cost of company-provided personal protective equipment.
- Plan for interim assessments to see if actions are moving forward. Upper management may ask for written progress reports.

A number of software programs are available for employers to use as a guide in developing plans. Many templates for procedures are available online.

A line manager or a safety coordinator need not take on the task of writing the plan alone. In some organizations, a safety committee performs the basic analysis of past activities and presents the annual plan for management review and approval. In smaller organizations, the operations manager, with help from a person assigned the role of safety coordinator, may draft a one-page safety program for the ensuing year. Loss control representatives from insurance companies may be able to assist but should not be asked to write the plan.

Look at the organization's needs and build on those. Where are the most injuries occurring? Have the safety teams focused on the hazards in those work areas and done all they can to eliminate or control the hazards? What OSHA standards apply to the business?

Federal and state OSHA offices offer free consulting assistance to employers. A telephone call to the area OSHA office will not encourage OSHA to make an inspection or an enforcement visit.

Another approach to safety planning and training

Some safety managers have found a calendar to be a helpful reminder or organizer of the many "must" items in a safety process and the training required to reinforce these processes. A calendar can be a convenient way to schedule safety activities and training because many safety events revolve around the months of the year. For example, February is the month in which employers are required to post the summary of the OSHA 300 Log in their facilities. The pleasant weather of the summer months makes them a good time to conduct a fire drill or mock evacuation of the facility. November and December often are times when safety planning committees meet to review the present year's activities and plan improvements for the coming year.

Temporary workers and their safety

The use of temporary workers is increasing, and safety coordinators need to be aware of employers' safety obligations to those workers. Safety coordinators should seek guidance from the company's human resources department or legal counsel so that the company knows its role in the training and hazard control activities for nonemployees who work at the facility.

Safety coordinators need to be familiar with several terms that describe the temporary worker and his or her work relationship:

- **Temporary agency employer**—the agency that sends the temporary worker to an employer; also known as the *primary employer* or *supplying employer*
- **Host employer**—the employer who assigns specific work to the temporary worker; also known as the *secondary employer* or *client employer*

Federal and state laws on temporary workers sometimes conflict, and some insurance regulations are unclear. A principal thread running through the laws and regulations is the question of who provides day-to-day supervision, assigns work to the temporary worker, and controls the work environment: the temporary agency or the host employer? In most cases, it is the host employer, and therefore safety and workers' compensation responsibilities cannot always be referred back to the temporary agency.

Definitions need to be made clear. Temporary workers are assigned by a temporary agency to an employer and report for work to and usually work under the general supervision of the host employer, although when large numbers of temporaries are employed, an agency may send an on-site manager to supervise their activities. Safety coordinators should check with risk management or the human resources department to ascertain how their insurance policy coverage affects temporary workers.

Safety coordinators should seek guidance from risk management staff to review general liability policies on the definition of "employee" because this definition varies among policies and among different states. Safety and security, access to the building, confidentiality, and trade secrets also need to be considered when temporary workers are being engaged.

Note that temporary workers are different from independent contractors, who offer their services to the general public, have a business office, invest in tools and equipment, and, as entrepreneurs, face a financial risk of profit and/or loss in their work. Independent contractors typically carry their own insurance coverage, including workers' compensation coverage. Good practice typically requires that independent contractors present a current insurance certificate as part of their contract requirements. Discussing safety responsibilities, worker safety training, and record keeping should also be part of contractual arrangements. As a general rule, if temporary workers are supervised by company employees, then they must be covered under the company's safety program. If they perform work solely under the direction of an independent contractor, they may be covered under the external safety program.

Safety training for temporary workers

Safety training should be a shared responsibility between the temporary agency and the host employer. Host employers should ask for a copy of an agency's training program and insist that referred temporary workers successfully complete such a program.

The host employer should arrange for integration of temporary workers into the employer's safety and health process, including specific orientation and training about workplace hazards, correct use of personal protective equipment, emergency evacuation procedures, and, where needed, medical monitoring. The general rule is that the host employer provides the needed personal protective equipment, unless the contract with the temporary agency requires the agency to supply personal protective equipment.

Job-specific training usually is the responsibility of the host employer, and the contract with the temporary agency should be reviewed to see how this responsibility is defined.

Recordkeeping responsibilities are not as clear. Under current OSHA regulations, the temporary agency is responsible for basic OSHA 300 Log record keeping, but the host employer also must record an injury if that employer has the day-to-day direction of the worker. The host employer can list recordable injuries to temporary workers on a separate OSHA 300 Log. But for the year-end log posting, injuries to temporary workers must be integrated with the totals for the employer's own staff. The host employer is responsible for notifying OSHA of serious injuries to temporary workers, such as a fatality or when three or more employees are hospitalized.

Some temporary agencies prohibit their workers from operating forklifts or other mechanical equipment, and this must be made clear in the contract review process. Temporary agencies also may want the right to inspect the host employer's facility.

Employers should establish definite communication channels with the temporary agency on the handling of workers' compensation cases, how to report job hazards, and what will be done about unsafe work behaviors.

Should the employer's first aid and medical staff treat work injuries and illnesses of temporary workers? Is there a liability aspect for the employer's medical staff? If a large number of temporary workers are employed, a medical clinic may need to be set up for temporary worker use.

International standards and safety

The workplace is increasingly becoming international. Emerging global markets have intensified the need for methods to standardize international safety regulations across many countries. The International Standards Organization (ISO) is a worldwide, Swiss-based organization of national standards bodies in 111 countries.

Two standards developed by the ISO are important for safety professionals to know: the ISO 9000 Series of quality management standards and the ISO Series 14000 standards, a series of international standards for environmental and safety management.

ISO 14000 is a series of voluntary international standards covering environmental management tools and systems. ISO 14001 recognizes that all management systems, whether for safety, quality, or environment, must provide a defined and organized approach to relevant activities while also meeting bottom-line business needs. These common elements include:

- Clearly defined policies and objectives
- Clear-cut responsibilities
- Documented systems
- Ongoing training
- Records and document control
- Control of critical processes
- Internal audits
- A method to correct problems
- Management reviews
- Continuous improvement

The ISO 14001 series does not establish performance requirements because it was written to apply to organizations of all types and sizes in diverse geographic, cultural, social, and economic situations. But the series does contain requirements that should result in a higher and steadily improving level of performance. The organization must meet the following:

- The requirements of its own policy
- A commitment to meet all applicable legislation and regulations
- A commitment to continual improvement of its environmental management system, which should result in environmental performance improvement

Coming next ...

The next chapter outlines the roles that orientation and job training can have in an effective safety and health process. Job orientation should include training in safe operating practices. Safe operating practices can be introduced successfully when they are an integral part of job instruction and the follow-up that is necessary for any business procedure.

9

TRAINING IN SAFE WORK PRACTICES

In order for all aspects of the safety program to be effective, a method needs to be developed to ensure that personnel are trained and competent to perform their safety management duties and ensure that communication of safety issues occurs among operational personnel and with the organization's management. Safety promotion plays a supporting, yet important, role in achieving effective control of safety risks during service delivery.

Safety promotion is a major component of a safety program. Together with the organization's safety policies and objectives, promotion is an important component of continuous safety improvement that is achieved mainly through the other aspects of safety management.

Safety promotion sets the tone for both individual and organizational behavior and fills in the blank spaces in an organization's policies, procedures, and processes, providing a sense of purpose to safety efforts. Through safety promotion, an organization adopts a culture that goes beyond merely avoiding accidents or reducing the number of incidents, although these are likely to be the most apparent measures of success. Rather, it leans more toward doing the right thing at the right time in response to normal and emergency situations.

Safety promotion supports safety culture communication, disseminates lessons learned, and enables the continuous improvement process.

Organizations should develop and maintain safety training programs that ensure that personnel are trained and competent to perform their safety program duties. Training programs should be adapted to fit the needs and complexity of the organization. The scope of the safety training should be appropriate to each individual's involvement in the safety program. The provision of appropriate training to all staff, regardless of their level in the organization, is an indication of management's commitment to an effective safety culture. The quality and effectiveness of training have a significant influence on the attitude and actual performance ("the professionalism") that employees will subsequently demonstrate in their everyday work.

Safety training and education should consist of the following:

- A documented process to identify training requirements
- A validation process that measures the effectiveness of the training
- Initial (general safety) job-specific training
- Indoctrination/initial training incorporating management systems, including human factors and organizational factors
- Recurrent safety training

Safe work practices are an integral part of core job training. In fact, more than 100 OSHA safety standards mandate some form of training to bring new or transferred employees into compliance. Other OSHA standards make it the employer's responsibility to limit certain job assignments to employees who have the special training to perform those particular job duties.

Figure 9.1. OSHA-Required Training Should Involve Adult Learners in Hands-On Activities and Group Problem Solving

(Goodluz/iStock/Thinkstock)

A Few OSHA-Required Training Topics

Emergency plans
Hazard communication
Fire prevention
Powered platform operations
Flammable and combustible liquids
Mixing vehicles
Process safety for highly hazardous chemicals
Hazardous waste operations
Respiratory protection
Accident prevention
Permit-required confined spaces
Control of hazardous energy (lockout/tagout)
Fire detection maintenance
Welding
First aid and CPR
Asbestos
Ionizing radiation
Lead
Laboratory safety
Cadmium
Bloodborne pathogens

Nonetheless, many employers hire and place new employees on the job with little or no job safety training. Conversely, organizations that value the importance of a safe workplace know that effective job training creates teamwork and morale, and so they support training with budgetary allowances. In so doing, employers and employees achieve a common goal: a productive and safer workplace.

So what is the best way to train employees on safe work practices? The employer should be sure to speak with the new employee about performing his or her job safely before the job starts, or on the first day of work. A job orientation and training process should include safe work practices as well as how to report unsafe conditions. Orientation should be followed by specific on-site training in the operational techniques and safe work practices of the job. This should not be an informal process. In fact, a formal safety orientation checklist should be developed.

Safety orientation responsibilities

A good trainer can be anyone within a company. Some companies choose a supervisor with a good reputation for training new employees, or someone who is competent in the job, or even a nonsupervisory employee who knows how to lead and motivate people. Perhaps these people could be the nucleus of a group of trainers who motivate workers to achieve excellence in production, quality, cost control, and safety.

Several groups or people are key to making a safety orientation program work: the safety office, which sets policy to ensure employees meet regulatory standards; the human resources or recruiting person, who orients the new employee to the company; the immediate supervisor and the job trainer; and the safety representative in the department where the employee will work. These people should all be aware of applicable regulations and company expectations in their training responsibilities.

In some cases, it may be necessary to have safety professionals with particular training or expertise and experience on a topic develop and conduct the training. Not only may the particular regulatory aspects of an OSHA topic be complicated, but the subject matter may also be extremely technical. Workers deserve to be trained by competent and qualified safety professionals with expertise on the subject matter, when necessary.

Human resources role

The training and orientation provided by the human resources department should cover the following general rules and procedures: the company's safety philosophy, the facility's safety plan and objectives, and the role of the safety committee. Copies of safety plans and policies should be provided to each employee.

Human resources staff also should emphasize the employee's responsibility to work safely and to observe the safety rules and regulations of the company and the job. New employees should review the company's general safety rules with the human resources staff, who should ask the employees to acknowledge in writing the receipt of these rules. New employees should also be told that safety rules are enforced and that all injuries must be reported.

Other topics to review with new employees include:

- Smoking/no smoking rules and areas
- Policies banning alcohol/drug possession and use on company property and policies banning possession of weapons or firearms on company property

- The organization's harassment-free workplace policy
- Special emergency evacuation procedures

Supervisor, job trainer role

More specific job-related safety regulations and procedures should be presented by the immediate supervisor, job trainer, or safety professional. Such training should include a description of the function of the department and the employee's particular job using a written job safety analysis and job description.

At this point in the training process, the supervisor or trainer should emphasize that maximum effort has been made to protect the employee through the purchase, installation, and use of safe equipment, machine guarding, and personal protective equipment. In addition, the company safety policy and facility and department safety goals should be restated.

The new employee should review and sign a statement acknowledging the receipt of a copy of the department's safety rules. In addition, the supervisor or trainer should point out and demonstrate potentially hazardous areas and jobs as well as review the standard of housekeeping in the department and the employee's role in workstation tidiness and housekeeping.

The supervisor or trainer should introduce any personal protective equipment required by the job. This includes:

- Providing equipment to the new employee
- Demonstrating how to wear personal protective equipment correctly and comfortably
- Indicating where to obtain replacement equipment
- Asking the employee to acknowledge receipt in writing

Safety officer/department training responsibilities

Perhaps the best way to ensure that all required safety training is being completed as required is to ask the safety officer or safety department. Not only should all required training records be maintained for all employees at one centralized location, but this centralization makes it easy for the safety department to review and document that all required training is taking place.

No other individual or group within the organization will be as familiar with all the required training as the safety office. Staff there should be the most aware of the requirements in all areas, and they also should be aware of all upcoming changes in industry standards on worker safety and any regulatory requirements that would warrant new or additional training on any given safety topic.

The new employee should review applicable Safety Data Sheets for the job and know the location of these sheets for future reference. In addition, the new employee should review the facility's written hazard communication program and know where a copy of that program is located.

All new employees should know the location of the medical department, backup first aid service for off-shift workers, eyewash stations, and emergency showers. They also should know that self-treatment of work injuries is not permitted and that medical and first aid services are provided for treatment.

In addition, new employees should know:

- Proper lifting procedures for the job
- Rules prohibiting temporary removal of machine guards
- How to report defective equipment to a mechanic or supervisor
- Location of areas for rest periods, eating, smoking, and nonsmoking
- Location of waste disposal and proper containers for paper, scrap materials, and used rags
- Location of fire extinguishers, facility policy on who should handle fires, and emergency evacuation routes from the work area
- Emergency machinery shutdown procedures and evacuation procedures

The supervisor or trainer should conduct a post-training test to assess the new worker's understanding of the training. Finally, the new employee should be asked to demonstrate the job. The old saying "If the trainee hasn't learned, then the teacher hasn't taught effectively" still stands.

To the greatest extent possible, the training should be conducted so that its organization and meaning are clear to the employee. Training should include: (1) overviews of the material to be learned; (2) information directly related to the workplace; (3) summaries and lists of key points. These tools will help ensure employee interest and retention of the materials.

In addition to organizing and creating the content, employers must also develop the structure and format of the training. The content for the program, the nature of the workplace or training site, and the resources available for training will help employers determine for themselves the necessary frequency of training activities, length of the sessions, instructional techniques, and individual(s) best qualified to present the information.

Employees must be convinced of the importance and relevance of the material in order to be motivated to pay attention and learn the material that the trainer is presenting. Motivation to learn the material can be improved by: (1)

clearly stating the goals and objectives of training; (2) relating the information to the interests, abilities, and experiences of the employees; (3) outlining the main points to be presented during the training session(s); and (4) describing the benefits of the training (e.g., the employees will be better informed, more skilled, more valuable both on the job and on the labor market, and at reduced risk).

An effective training program allows employees to participate in the training process and to practice their skills or knowledge. This will help to ensure that they are learning the required knowledge or skills and will permit correction if necessary. Employees can become involved in the training process by participating in discussions, asking questions, contributing their knowledge and expertise, learning through hands-on experiences, and learning through role-playing exercises.

Department supervisor follow-up

The new employee's supervisor should conduct follow-up refresher training a few weeks after the orientation training. During the refresher training, the supervisor should:

- Observe whether the new employee followed job training correctly
- Evaluate whether the job is being performed efficiently and safely and meets quality and quantity requirements; if not, determine what retraining is needed
- Determine how much additional time is required to bring the employee up to the full standard of performance
- Ask the employee what was most important about the training received and whether he or she found it easy to follow the safety regulations and procedures

Effective learning for adults

Boring orientation and training sessions lead to daydreaming and poor information retention. The following recommendations will help keep employee attention on the training:

- The atmosphere of the training room should be relaxed and informal, with coffee or soft drinks to set a positive and friendly tone.
- The business objectives of the training should be clearly outlined at the start, and the training should be specific to the workplace.
- Several employees should be trained at the same time so that they can learn from one another's questions and experiences.
- In today's diverse, multilingual workforce, job orientation and training in languages other than English may be necessary.

Some organizations use interactive intranet training to provide much of this basic orientation and training information. This type of training allows the new worker to proceed at his or her own pace and prevents advancement to more complex procedures until the earlier step or information has been mastered. Such training also automatically records who has completed the training and their acknowledgment that they have read and understood company and department work rules. But the immediate supervisor's personal involvement during job training remains key to long-term, effective, safe work performance because supervisors are the leaders to whom new employees turn with questions and concerns.

Workers who have had good learning experiences are more likely to take the learned message back to the job and apply it.

OSHA and safety training

OSHA requires employers to train employees in OSHA safety standards and ensure that employees can apply the standards on the job. In fact, OSHA considers safety training one of the cornerstones of an effective safety process. According to OSHA, an effective safety process includes management support, employee involvement, worksite analysis, hazard evaluation and control, and safety and health training.

Employee training often is an issue after an OSHA compliance officer has cited an employer for violations of safety standards. The absence of employee training or the inadequacy of the training to meet requirements of the standards may be at issue when management must defend its actions in an OSHA hearing.

Coming next ...

The next chapter will deal with the role of safety and health inspections and the important thought process that should be conducted before initiating such activities.

10

SAFETY AND HEALTH INSPECTIONS

The purposes of workplace safety and health inspections are to discover hazards and unsafe work practices before injuries occur and, through the recommendation and correction process, to eliminate or control those hazards and eliminate or correct unsafe work practices.

Preplanning for safety and health inspections

Safety and health inspections should not be undertaken unless management is fully prepared to correct the unsafe conditions and work practices uncovered by the inspections, including setting aside money and resources to do so.

An effective safety inspection activity also will help managers, supervisors, and the safety committee become more knowledgeable about hazards. This increased awareness should allow these hazard fighters to spot hazards more quickly and correct dangerous situations on the spot. Safety inspections also provide an opportunity to learn more about operational processes and to involve more employees in the safety program.

A well-planned safety and health inspection will provide a consistent method of recording observations, will use inspection time more efficiently, and will reduce the possibility that important items may get overlooked. People who know the jobs best, such as supervisors and shop floor employees, can help develop a format focused on potentially unsafe conditions and unsafe work practices.

Looking at unsafe conditions and work practices

When managers and workers think of developing a safety and health inspection program, often what comes to mind is the correction of hazards, dangerous equipment, and other unsafe work conditions. That is a key part of the program but is not all of the program. Safety experts say that as many as 90 percent of work injuries are caused by unsafe acts.

During the 20th century, safety experts reduced the reasons for incidents to two basic causes: unsafe conditions and unsafe job practices or acts. Typical examples of these two types of incident causes include the following:

Unsafe conditions

- Defective equipment
- Unguarded equipment or machinery or guards broken or not in place
- Cluttered, unstable, slippery floors
- Inadequate area lighting or task lighting
- Unstable loads of raw materials, in-process goods, or finished products
- Frayed electrical wiring or ungrounded electrical devices
- Environmental conditions such as noise, heat, dust, fumes, and vibrations
- Inadequate ventilation
- Improper storage of hazardous materials

Unsafe work practices

- Workers untrained in safe work practices for the particular job
- Safe work practices not being followed
- Operating equipment at unsafe speeds
- Removal of machinery safeguards and safety devices
- Unsafe lifting or materials-handling techniques
- Unauthorized use of equipment
- Sloppy work habits and poor work area housekeeping
- Failure to use provided and designated personal protective equipment
- Improper dress for the conditions of the particular job
- Horseplay, workplace antics
- Use of defective tools

Although many incidents are the result of an employee deviating from standard job procedures, doing something prohibited, or failing to do something that should have been done, safety coordinators should check to see if something in the work conditions led to or contributed to the unsafe practice. Some employees will take risks as a shortcut, and in some organizations, risk-taking is part of the company culture and is thought to improve efficiency or productivity. But the greatest concern is workers who don't understand the risks associated with their jobs or how to work safely to avoid those risks. Old work habits die hard—particularly if those techniques have worked, incident-free, for a long time.

Important steps in hazard detection and control involve distinguishing between worker error and supervisory or management system error and then focusing on what caused the system to break down. Human error is reduced when:

- Supervisors and workers know the correct methods and procedures to perform assigned tasks
- Workers are formally trained in the correct and safe work procedures of the job and demonstrate a skill proficiency before being left alone to operate a particular piece of equipment
- The organization gives priority to potentially dangerous situations and provides the corrective action necessary to avoid injuries
- Supervisors provide proper training, direction, and overview of workers' job performances; are cognizant of workers' skill levels with each piece of equipment; and accordingly adjust the supervision of each worker and any necessary retraining
- Employees are involved in developing safe work practices, which provides a more solid foundation for safe working routines as buy-in evolves during the development of safe work rules

The inspection format

The format of a safety inspection should list the specific defect or unsafe act reported and the corrective action recommended. Specific questions likely will result in more focused recommendations. Supportive paperwork should show how recommendations were handled and brought to a conclusion, such as being prioritized for future action and then resolved. In some cases, after investigation an item will be classified as "not really a hazard."

Most management groups take pride in correcting hazards and unsafe work practices themselves rather than having outsiders such as an OSHA compliance officer note them.

Creating an effective inspection format focusing on both unsafe conditions and unsafe acts need not be a laborious process. Ingredients for the form may be already developed and available, such as:

- Results of the hazard identification activities
- Fire inspection and protection reports
- The facility's emergency evacuation plan
- Safety Data Sheets for chemicals and hazardous materials used at the facility
- Reports from the insurance carrier's loss control consultant, who completes safety tours of the facility
- OSHA inspection and citation reports
- Local programs or checklists instituted to meet regulatory requirements (lockout/tagout) procedures, hearing conservation rules, and respirator training and enforcement)
- Written work procedures or facility job safety analyses that list potential hazards and the safe work procedures to overcome those hazards
- Informal comments from employees, such as "The flooring at the entry to the warehouse needs to be repaired" or "It's dangerous working near those old scrolling machines."

Although it is important to correct deficiencies, the format also should include areas for positive comments. These comments could be "All department personnel wearing hearing protection effectively—congratulations" or "Fork truck operators sound horn approaching warehouse intersections." Some formats provide for grading per criteria, such as "excellent," "good," "needs more work," or "refresher training required."

Who should inspect?

Some organizations ask safety committee members or department safety representatives to conduct inspections. Other organizations find it effective to mix supervisors and safety committee members in the process. Some organizations have one person conduct the inspection; others insist that two people work as a team. Whatever approach is most effective, members should be trained in the company procedure of safety inspection and in what exactly they will be asked to inspect.

A key ingredient to a successful inspection is involvement of and participation by line managers and the supervisors who are responsible for safety in their departments. Supervisors may be knowledgeable in one department or one specific area but may lack in-depth knowledge about other areas of the facility. Pairing the supervisor with an experienced employee may provide the balance in perspectives needed.

Management members with wide-ranging responsibilities, such as production managers or superintendents, by the very nature of their positions can bring broad perspectives to a safety inspection. They also can be teachers and trainers of less experienced inspection personnel.

What to inspect?

In smaller facilities, a safety and health inspection can cover the whole facility in one tour. But as the complexity of operations or service work increases and as the number of employees and the size of the facility rise, it may not be possible to conduct a thorough inspection on each tour. In such situations, managers should limit the tour to a reasonable number of departments where in-depth inspections can take place. Other managers may alternate full general tours with focused inspection tours that emphasize specific items such as potential electrical hazards, ladder and scaffold safety, or employee work practices when handling chemicals.

How often to inspect?

The traditional monthly safety inspection tour is inherent to the safety activities of many organizations. Other organizations schedule weekly inspections focusing on specific areas of the facility or highlighting one element of the safety program. Tours scheduled more frequently than once a month will markedly reduce employee exposure to hazards and provide more lead time to correct unsafe work practices.

Some organizations allow an hour for these tours; other organizations allow more time. Many organizations field-test the inspection to determine what they can accomplish in a reasonable amount of time.

Types of safety and health inspections

Safety and health inspections can be conducted in many ways, but two approaches are used most frequently: the general facility inspection and a specific department or hazard detection inspection. The former is usually thought of as a broad overview of general hazard potential and work practices in the facility and is perhaps the best way to kick off an inspection program.

Specific safety inspections tend to deal with exactly that—the specifics—

either by department or by function, and cover subjects in depth such as the wearing of hearing protection, reviewing safety orientation practices for new hires, or checking the outside storage of propane tanks.

In sophisticated organizations, safety inspections may fall under the auspices of the quality assurance program. Quality assurance programs may be described simply as three main functions that involve the evaluation of safety and associated programs or systems. These functions are surveillance, audits, and program reviews.

Surveillance activities include periodic and routine inspections of a variety of operational or functional areas. For example, chemical storage tanks may be checked by visual inspection daily to ensure that the correct levels are maintained and as a double check of gauges and level indicators. These may also include such activities as a daily walk-through of the plant to ensure that machine guards are in place or that workers are wearing PPE effectively.

Audits are more comprehensive assessments of the plant and its operations and provide a "snapshot" of the safety conditions in the facility at a given time. Audits are planned inspections that are typically scheduled and cover specific activities within a department. A safety audit at a large facility may cover the confined space and respiratory protection programs one year and the hazard communication and hearing conservation programs the following year. Audits are typically conducted by professionals with particular industry expertise in a given area. In addition, audits may include a review of records and corrective actions, observation of worker activities, inspection of equipment, and evaluation of administrative programs.

Finally, program reviews are the third tier of quality assurance means to evaluate safety programs. A comprehensive program review looks at the entire safety program and all associated processes to determine the level of adherence to all applicable regulations and industry standards.

Prioritizing recommendations

Safety and health inspections, audits, and reviews will generate all sorts of recommendations. A few involve emergency situations that could endanger workers and must be corrected immediately. Other observations should be placed on a descending scale of importance so that they can be handled by the department involved or through maintenance scheduling. High-cost changes and improvements requiring capital funds will need review and approval from higher levels of management.

Some organizations ask inspectors to use one of the following rating scales to indicate the importance of the particular hazard or unsafe act uncovered:

- Immediate (I), serious (S), scheduled for maintenance repair (M)
- Priority listings of 1, 2, 3, or 4, with 1 being the most important and requiring immediate attention
- Red stickers (needs immediate attention) or orange stickers (needs priority attention) to mark important items uncovered

Computer software allows inspectors to immediately convert their audit findings in each department to a record of what was found and the inspectors' recommendations on priorities and corrective steps. Department managers can then add their corrective actions to the report, providing a continuing record of what was found and what was corrected in what time frames.

Follow-up

Be sure that the inspection activity is tied in with action steps for follow-up. Assigning responsibilities in terms of follow-up is critical to the usefulness of an inspection activity. Some organizations require the department manager or supervisor to report within three days on the corrective action taken, with the clear directive that serious, urgent, or high-priority recommendations should receive immediate attention. In other facilities, the supervisor and the safety committee member address the items needing attention, and the safety committee member then reports on the results at the regular safety committee meeting. Some recommendations for improvement may require lengthy program development or the purchase and installation of special safety systems or equipment.

Some items will require capital funds or long-term action in order to be corrected. Progressive management organizations frequently establish a management steering committee (as outlined in Chapter 7, "Involving employees in injury prevention") to oversee the inspection activity and ensure that items do not fall through the cracks.

Coming next ...

The next chapter will deal with injury or incident investigation—finding out what went wrong when an incident occurred and how an investigation, handled thoroughly and promptly, can do much to decrease workplace hazards that cause incidents.

11

INCIDENT INVESTIGATION

When a company reacts swiftly and positively to serious incidents, its actions speak loudly about its commitment to the safety and health of its employees. A thorough injury investigation procedure is the essential link between events showing where a safety process falls short and the actions needed to improve that process.

For a variety of reasons, the terminology of accidents, incidents, and near misses has become a bit convoluted over the years. For many, the term *accidents* has a connotation of "unavoidable or unpreventable," as in a lightning strike. There is the notion that since you don't know when lightning is going to strike, there is nothing you can do to prevent it from happening. There is also the notion that in order for an accident to occur, there needs to be an unsafe act or condition that leads to the undesirable outcome.

These notions have two significant flaws, however. Even a lightning strike is not unavoidable or unpreventable. For example, if there is a thunderstorm brewing, you can prevent yourself from being struck by lightning by not going out on the golf course. If your car has a flat tire on the highway and it causes you to swerve into a parked car, it may not have been preventable, but it wasn't necessarily an unsafe act or condition that caused you to hit the car.

Therefore, the term *accident* in this book will refer to undesirable conditions and acts that lead to events that result in adverse outcomes such as injuries, illnesses, and property or equipment damage.

The term *incident* will be used to describe undesirable conditions and acts that have the *potential* to cause adverse outcomes such as injuries, illnesses, and property or equipment damage. An example of an incident could be a worker forgetting to replace a guard over a hazardous part of a machine. The failure is observed and corrected before anyone is hurt, but the potential for harm was there and the "incident" warrants investigation and corrective action.

The term *near miss* will be used to describe an event that resulted from undesirable conditions and acts, but in the near-miss situation, those conditions and acts just barely avoided a significant event where injury or property damage could have occurred.

From the discussion above, it should be apparent that both accidents and near misses can be considered "incidents." For that reason, the term *incident* is often used pervasively in numerous industries and will be used throughout this chapter.

Steps to take following an incident

Confusion often follows an incident, with co-workers, supervisors, and medical personnel wanting to help. Because sorting out the confusion can be difficult at times, a prearranged procedure or drill will help supervisors get important matters taken care of promptly.

The following suggestions will help formalize a procedure of proactive steps to take following an incident:

- Arrange for prompt first aid or medical treatment for the injured worker. If the employee needs to be transported to a hospital, accompany the employee in the ambulance, or have a dependable co-worker go with the employee.
- Within the framework of company policy, notify the family of the injury and the highlights of workers' compensation and other benefits to which the injured worker will be entitled. If the employee came to work in his or her own car, arrange to have the car driven to the employee's home.
- Avoid adversarial relationships with workers by telling them up front what they are entitled to. Review workers' compensation entitlements with the injured worker (in some organizations, this may be handled by the human resources department) and the likely course of medical treatment. Advise the injured worker of the name of the insurance carrier, the contact person at the carrier, and the telephone number to call. Assure the injured employee that the supervisor and the human resources department are available to assist with any questions or concerns.
- Conduct an investigation of the incident to find out what went contrary to company policies as well as to determine any unsafe conditions or unsafe work practices. Interview any witnesses independently. Interview the injured

worker when that person is able to offer a coherent account of what happened. Ask the injured worker for recommendations to prevent a recurrence. Prepare the organization's incident investigation report.

- Implement the findings of the investigation. If higher-level approvals are required, promptly bring them into the discussion and planning.
- Keep in touch with the injured worker (at least weekly by telephone or personal visit), and assess any problems with workers' compensation payments. Deal with any problems.
- Contact the treating physician to explain the nature of the employee's job and the company's willingness to have the employee return to modified work on a temporary basis as soon as possible.
- Welcome back the returning injured worker and, when necessary, assign that person to a job suitable for any work restrictions.
- Follow up to confirm that return-to-work progress is continuing.

Why conduct formal incident investigations?

Good hazard investigation work will reveal gaps or failures—holes that need to be plugged by improved hazard identification, elimination, or control; job retraining; or perhaps more effective use of machine guarding or personal protective equipment. A formal incident investigation procedure, supervisors trained in its use, and management follow-up can provide the clues to deficiencies that, when corrected, can reduce the recurrence of injuries.

Incident investigation is more than filling out a company form. An investigation procedure should identify:

- Who is to be involved in the investigation process
- Correct preparation of a factual report for management
- Initial steps to take to prevent recurrence of the incident
- Long-term actions that may be necessary to prevent further injury and the person responsible for each action
- A provision for training or retraining of employees if an unsafe work procedure was the cause
- Reaffirmation or revisions to safe operating procedures

Supervisors should be trained in the company's requirements of incident investigation and reporting. Requiring reporting forms to be filled out completely is key to keeping a supervisor involved in incident prevention activities. Assign investigation and reporting responsibilities to the immediate supervisor, with overview by management and, in some cases, by the facility safety committee.

The format of a useful incident investigation report should include the following:

- Accurate information about the employee, his or her work experience and job tenure, and the nature of the injury
- The employee's signed statement of the sequence of events resulting in the incident
- Additional information about what happened that is gathered from witnesses, co-workers, or a supervisor
- Identification of immediate causes of the injury and any secondary or contributing causes that created the incident situation

These initial investigation steps logically lead to determining the stages of corrective action, including:

- Corrective action that already had been taken by the supervisor by the time the written report was prepared. A 48-hour time period typically is allowed in which to get the initial report completed and into the hands of the first reviewer.
- Recommendations for additional corrective actions that relate to the specific injury and/or the overall operating system and safety process. Review and approval by one or more levels of management often occur.
- Consideration of these questions: What went wrong in the existing management process that resulted in the incident? What needs to be fixed in the present management process that allowed the incident to occur?

Who should investigate incidents?

An important early consideration in planning incident investigation activities is who will be involved in the investigation. At the start, the injured worker and his or her immediate supervisor should be involved. In many organizations, the immediate supervisor conducts the investigation by speaking with the injured worker. The supervisor then signs a completed investigation form and sends it to a superior. In other settings, one or two additional people, perhaps a safety committee member and a safety coordinator, assist the supervisor in conducting the investigation and preparing a report. In some organizations, many people are involved, including safety committee members or union representatives and higher levels of management. In complex incident situations, professionals such as an engineering manager, an industrial hygienist, or an occupational health physician may be asked to help determine the cause(s) of the injury or occupational illness.

The immediate supervisor should be a key person in all incident investigations. Supervisors know job procedures, hazards, environmental conditions, and unusual circumstances that may be encountered on a job. They know their employees' job experiences as well as each employee's personal strengths and weaknesses.

Investigating injuries also increases the supervisor's sense of responsibility for incident prevention. The immediate supervisor has the primary responsibility for implementing corrective actions to prevent recurrences, but it should be noted that in major incident investigations, there are drawbacks to having the supervisor handle the investigation alone. The outcome may be influenced by that supervisor's experiences, opinions, attitudes, and self-interests. Also, an immediate supervisor usually does not have broad experience in machine design and equipment installation or management policies and approval levels, expertise that may be needed in some investigations.

At a later time, other professionals may be involved in the injury investigation process. An OSHA compliance officer may receive a complaint about the circumstances surrounding the injury and come to the facility to investigate. A state labor inspector may be involved, or representatives from the employer's workers' compensation carrier or loss prevention service may review the case and the steps taken to prevent recurrence.

In the case of a work-related fatality, company management and legal counsel should be contacted immediately, and the incident should be investigated by a team that includes the first-line supervisor, senior management personnel, engineering staff, and legal counsel. In the case of a work-related fatality or the hospitalization of three or more employees, the incident must be reported to OSHA *within eight hours*.

Questions to ask

Organizations approach incident investigation in different ways, depending on their company policies or cultures and the nature of the work processes, but a structured approach to incident investigation is needed. Learning the where, why, when, how, and to whom injuries are happening means being better able to prevent them in the future. The following are important questions to be raised in an incident investigation:

- Who was injured?
- Who witnessed the incident, and what did these witnesses see, hear, or feel when the injury occurred?
- What did the injured employee say happened, and what job function was being performed?
- What were the environmental conditions (air, heat, humidity, cold, ventilation) of the work area when the incident occurred?
- What, if anything, was not normal before or after the incident?
- What were the mechanical, electrical, and housekeeping conditions of the equipment involved?
- What lighting was provided in the work area?
- What training on the job had the injured worker received?
- What materials, machines, or tools were being used?
- What personal protective equipment, if any, was required on the job, and was the worker wearing it?
- What machine guards, if applicable, were in place or not in place?
- What work practices were being followed? Were they safe work practices?
- What safety rules were not followed?
- What unsafe operating conditions, if any, were involved?
- What should be done to prevent recurrence?
- What will be done, by whom, and by what dates to prevent recurrence? What new or revised work procedures, if any, will be needed to prevent recurrence?
- What elements of the management system of hazard identification and control need to be reexamined or reaffirmed with applicable workers?
- When did the injury occur?
- When was the injury reported?
- When did the employee start with the company and start this job?
- When did the employee receive job safety training for this job?
- Where in the facility did the injury occur?
- Where were the witnesses located at the time?
- Where did the injured employee first receive first aid/medical attention?
- Where on the body was the employee injured?
- Why did he/she do what he/she did?
- Why was the employee in the position that he/she was?
- How did the employee get injured?
- How many times had the injured worker performed the particular task associated with the injury?
- How could the employee have avoided the injury?
- How could fellow workers or the supervisor have prevented the injury? How could company management have prevented the injury?

Incident investigation supplies

Not every injury scene needs to be recorded or preserved, but whether or not to do so is an important consideration when a complex incident is being examined. This is particularly the case when:

- Serious injuries or fatalities are involved
- Substantial property damage occurs
- A suspicion of sabotage or arson exists

Supervisors should have the following materials available to assist in the incident investigation process:

- A current copy of the company's written procedure for incident investigation
- Company incident reporting forms
- In the case of a serious injury or building or product damage, yellow tape to cordon off the incident area immediately after the injured person has been removed for medical treatment but before normal work resumes

Video footage or photos can document many facts that may otherwise be lost in the confusion and concern for the injured worker and can help recall information, perhaps years later, for a court case. Both long-range and close-up shots of the injury scene should be taken. Use of a portable tape recorder allows an investigator to dictate notes as conditions are observed, saving time in having to record details by hand.

OSHA requires employers to advise the agency within eight hours of a job-related fatality or an incident where three or more workers are hospitalized as a result of the same incident. In the case of a workplace fatality, local police also should be advised.

In serious injury cases, the workers' compensation carrier should be notified by telephone because that organization may want to conduct its own investigation. In the case of property or product damage, the company's insurance carrier also should be notified.

Interviewing witnesses

Obtaining accurate information from the injured employee and those who witnessed the incident is important in order to determine the cause(s) of serious injuries. The accuracy of witnesses' memories will be highest immediately after the incident. Witnesses should be interviewed privately and separately—if witnesses are allowed to converse and compare accounts, recalled details may be compromised, further delaying the investigation into what actually happened.

Investigators should try to put the witness at ease at the start of the interview and assure him or her that the aim of the investigation is only to discover the cause of the incident. Investigators should express concern about the injured employee and emphasize that information is being sought to prevent recurrence of the injury. Let the witness talk. Ask for suggestions from the witnesses to prevent recurrence. If a witness's statement is unclear, recite the statement back to the person and ask for clarification.

Investigators should take notes and review the statement with the witness at the end of the interview to make sure the investigator has a correct summary of what happened. Some organizations require two incident investigation forms to be completed, one by the supervisor and another one by the injured worker.

Instead of jumping to conclusions, investigators should be patient and wait until all information is in. The investigator should continue probing opinions and statements of those involved until the investigator believes fact gathering is complete.

Why did the incident occur?

Incident investigations work in unpleasant territory, trying to find out what happened and to get at the root causes of the incident. Workers may view the formal incident investigation as a faultfinding process. For fear of discipline or other repercussions, witnesses and even injured workers are at times reluctant to give the full story of what happened.

Many incident investigations amass statements such as "Employee was careless" or "Employee did not follow safety instructions." The emphasis in incident investigations should be on determining what went wrong with established safety procedures and what must be done to get the work system back on track. Perhaps procedures are hard to follow and need to be reexamined.

Do workers intend to injure themselves? Not usually. Thus, investigations need to uncover the genuine reasons for the injury, not just the factors that led to it. Frequently, injuries are due to a lack of training or regular reinforcement of that training, as well as a misdirected motivation to get the job done regardless of safety considerations—a work culture that may be fostered in the organization.

A worker should not be blamed or penalized for an incident over which he or she had little control. Workers do not design machine guarding systems, but they should be trained how to use the guarding systems effectively.

What happens after the investigation?

The most important elements of incident prevention are the steps that follow the investigation. The supervisor may have taken immediate action to stop the equipment and get the unsafe conditions corrected. But the most serious incident situations usually involve longer-term steps and frequently involve higher levels of management and allocation of funds and resources beyond the scope of authority of the immediate supervisor.

Three key elements are involved in incident investigation follow-up:

1. Who will do what?
2. When will the corrective actions be completed?
3. What was the cost of the incident—medical and workers' compensation costs, machinery downtime and effects on quality, replacement costs of personnel, and/or damage to equipment and tools?

Organizations with effective incident prevention procedures require review of the written incident prevention report by higher levels of management or by a management safety steering committee or a joint safety committee. The goal of this review is to ensure that the follow-up and prevention activities that will be undertaken are designed to prevent recurrence and that the prevention activities actually take place.

Senior managers who will review incident reports should insist that all questions on the incident investigation form be answered and that the recommendations to prevent recurrence are targeted, with responsibility assigned and completion dates set. They also should insist that recommendations to prevent recurrence be specific and thorough.

Administrative procedures should be in place to record incidents and develop appropriate corrective actions to prevent recurrence of similar events. Follow-up procedures should track and trend corrective actions to ensure that they are fully completed and are done according to schedule.

Supervisors should undergo short training sessions to ensure that they understand the incident investigation process, the forms involved, and their particular roles in injury investigation as determined by company management. Safety personnel, human resources staff, and nursing personnel should be identified as resources for line managers to turn to for assistance, but the responsibility to investigate the incident should be the supervisor's.

Investigating "near-miss" incidents

Most organizations investigate incidents involving an employee injury or a job-related illness. That is a solid start, but it may not be enough.

Forward-thinking organizations with successful safety processes go further—they investigate any incident where no one was injured but someone could have been: the "near misses." Near misses with serious injury potential could include incidents such as the following:

- An office employee leans back on a chair and topples backward. The employee is dazed but not injured and returns to work. However, a serious head injury could have occurred. Thus, an investigation needs to be made to determine whether the chair is defective or the employee's actions were responsible. Also to be looked into is whether other workers do the same thing.
- A trades helper assisting painters in setting up a scaffold falls from the scaffold's lowest level. The helper is not injured and returns to work, but the incident could have resulted in a broken leg or worse. Did an unsafe condition with the scaffold exist or do work practices need to be reexamined?
- A forklift operator bumps a storage rack, and a bag of concrete mix falls 20 feet and just misses another worker standing a foot away.

Near misses send a message: "Next time, you may not be so fortunate." Like work injuries, near misses call for quick investigation and action before the situation repeats itself—but with more serious consequences.

The investigation team may include building or facilities managers, in the case of property damage, or materials managers, in the case of raw materials or product damage. In the case of property damage, insurance investigators from the property insurance company frequently come on site to conduct their own investigations.

The conclusion is clear: Management should heed the warnings of near misses so that workers won't be injured by a real incident.

Coming next ...

The next chapter will describe the core programs and systems that need to be in place in order to effectively respond to emergencies and disasters.

12

IN THE EVENT OF EMERGENCY ...

As mentioned in the previous chapters, most businesses have several basic activities that help control hazards, prevent potential injuries, and reduce costs. While not directly in the realm of safety, they are important and necessary building blocks of an effective safety and health process.

For example, procedures usually are in place for first aid or medical service, emergency preparedness, and workplace violence. In addition, most companies have policies for dealing with alcohol and drug problems. These activities may be coordinated by a human resources manager, a risk management professional, an engineering or fire prevention officer, a security specialist, an administrative manager, or a facility nurse.

Regardless of which department coordinates these activities, all come together as part of the safety process.

Medical/first aid services

For small offices, a basic first aid kit may be all that is needed to treat minor cuts and simple health problems. For larger businesses with complex and potentially hazardous operations, broader coverage is needed, including trained first aid attendants, occupational health nurses, medical staff, and readily available ambulance services.

OSHA requires the "ready availability of medical personnel," such as a clinic or hospital emergency room reachable within four minutes of a business in case a serious injury with severe bleeding should occur; otherwise, trained first aid personnel are required on site.

In business activities involving hazards, having trained medical or first aid persons on each shift will result in immediate cost savings. These attendants can treat minor injuries or job illnesses without the cost and lost time involved in sending injured employees to a medical facility. Self-treatment of injuries by employees should not be allowed, however.

First aid staff and supervisor training is critical in an emergency. An employer can easily be embarrassed in front of workers and in the public press because of poor staff training in handling emergencies.

The following basic first aid and medical services should be in place:

- Medical personnel should be available for advice and consultations.
- First aid personnel should be formally trained and certified in emergency first aid procedures, CPR, the Heimlich maneuver (to release throat blockages in choking victims), and in details of the OSHA Bloodborne Pathogens standard that applies to medical and first aid attendants assigned by the employer to render first aid.
- Contact numbers of emergency medical service, fire departments, police, and local clinics and hospitals should be posted and available to medical personnel, first aid personnel, and supervisors.
- Emergency facilities for eye washing and skin decontamination should be present at locations where corrosive materials are stored or utilized.
- A local company physician or clinic with training in occupational medicine should be selected and should provide the employer with written standing orders regarding initial or first aid treatment of injured or ill workers and with medical and first aid supplies on hand.
- A first aid log should be kept of every injury or job illness reported. Data should include employee name, department, date and time of the injury, the extent of the injury, the treatment applied, any follow-up required, and the name or initials of the first aid or medical person who treated the injury.

If further medical treatment is required, arrangements need to be made for a co-worker, taxi, or ambulance to transport the injured employee to a clinic or medical facility. The relatively small cost involved in this type of training and preparatory work will pay off many times over in a staff that is confident in handling emergency situations and that demonstrates their effectiveness in doing so. Employees trained in first aid have a heightened safety consciousness and significantly lower incident rates. Trained first aid people add to the overall safety consciousness of the facility.

Are your workers reluctant to report injuries or exposures?

The problem sounds strange: An employer provides medical and/or first aid services to treat job injuries, and yet a small number of employees prefer to treat the injuries themselves or let the injuries go untreated.

The vast majority of employees report incidents and have injuries treated at the job site. But some employees are embarrassed to report that they have had an injury. Some workers are reluctant to report injuries because of worries about investigations, excessive paperwork, or disciplinary actions. Age, race, gender, and ethnicity are demographic factors that may affect a person's willingness to report an injury or seek help.

Management communication efforts on safety also may be a cause of nonreporting. In many organizations, injury and illness statistics propel the safety program. Charts show the number of lost-time injuries and trends of injuries. And some employers take the position that unless an injury is reported promptly for treatment, they will pressure their insurer to deny the workers' compensation claim.

To maximize work injury reporting and treatment, an organization needs to develop a work culture in which workers feel comfortable enough to report incidents without fear of retribution. The organization develops this culture by providing a safe working environment and demonstrating a positive management attitude that shows concern about workers' safety. Management also needs to recognize that some types of occupational health problems—such as dermatitis, industrial asthma, or work-induced hearing loss—may take years to develop into a full-fledged work-associated problem.

Employers that provide medical and first aid services want their workers to use those facilities for treatment of injuries and job-related health problems.

The following suggestions can encourage the use of in-company services:

- Offer a clean, neat first aid facility with a staff that works helpfully and confidently with injured workers. The process also needs to be swift and efficient.
- Ensure that employee training describes possible signs and symptoms of injury or exposure.
- Make sure training emphasizes the importance of timely reporting and treating, through company facilities, of injuries and exposures to chemicals, infectious agents, or other hazardous materials.
- Make sure follow-up occurs to ensure that the worker is following medical treatment regimens.
- Counsel those who are reluctant to report injuries and determine the reason for the reluctance.

A positive atmosphere actually may increase incident reporting for a while, but in time, injury frequency will moderate as the workplace becomes safer.

Emergency preparedness

Workplace emergencies and violence are facts of life today. It is important to train employees in emergency procedures and evacuation processes in the event of fire, explosion, severe storm, bomb threat, or workplace violence. Inadequate or nonexistent emergency plans can turn incidents into real disasters, with the potential for loss of life, product and building damage, and adverse publicity.

While some natural disasters can't be prevented, good planning and practice can limit the extent of damage and save lives. Emergency planning allows key personnel to manage the problem so that it does not spread and to decrease the chance of injury to employees and prevent panic, which can make a bad situation worse. Key personnel are far more likely to carry out emergency measures on an orderly basis if they are properly trained, and the training can be reinforced through tabletop exercises, drills, and real-life exercises and mock evacuations.

Managers and supervisors should understand clearly their specific responsibilities in an emergency situation, how to direct the orderly and safe shutdown of equipment, and the predetermined routes to evacuate employees from the facility. Employees need to know clearly what to do if an emergency is declared in their facility. Assembly points outside the facility and away from the building should be clearly indicated for each department, and arrangements should be made to account for all staff following the evacuation of the facility. Supervisors may be assisted by a department fire warden or emergency coordinator.

Organizations that are tenants of a public building should check with the building manager and integrate their activities with the overall building security, emergency preparedness, and evacuation plans.

Federal agencies have established regulations explicitly requiring employers to have emergency action plans in place for their workplaces. These regulations are a sound basis from which to develop procedures for emergencies, security, and violence control. For example, the regulations stipulate that employers should appoint an emergency coordinator at the management level with responsibilities to:

- Determine whether a situation exists that requires activating emergency procedures
- Direct all efforts in the area, including evacuating personnel and minimizing property loss
- Authorize the call to outside emergency services, such as community fire departments and medical aid
- Direct the shutdown of business operations when necessary

Security checklist

A security policy and procedures checklist should include the following:

- Arrangements for a professional security audit of the facility
- Security badges with photographs worn at all times on the premises
- Review of procedures by the risk management department and the company's insurance carrier
- A control system for the building and office keys
- Supervisor training on how to handle security issues
- Employee training that emphasizes not to put themselves or others at personal risk
- Evaluation of local police capabilities in the event of an emergency
- Provisions for armored car or a security service to transport money to banks
- Provisions for locks on all exterior doors that permit exit but prevent entrance
- Regular change of computer passwords after employee terminations
- Regular backup of key computer data
- Fireproof storage of computer records and other key information

The regulations also require employers with 11 or more employees to develop written emergency action plans that include emergency escape procedures and routes and to post that plan for employee review. In addition, the regulations require:

- Training to assist in the safe and orderly evacuation of employees
- Developing emergency procedures to be followed by employees who remain to operate and shut down critical operations before they leave the facility
- Creating a procedure to account for all employees after emergency evacuation has been completed
- Assigning rescue and medical duties for employees
- Implementing policy on the designated procedures of reporting fires and other emergencies
- Listing names and job titles of persons who can be contacted for further information or duties under the plan
- Marking exits with clearly visible signs
- Ensuring that exit doors serving more than 50 people swing in the direction of the exit travel
- Ensuring that no locks or fasteners prevent free exit, except in the case of psychiatric or corrective institutions where supervisory personnel are in attendance at all times

- Providing fire alarms to warn occupants about the existence of fire
- Marking nonexit doors or passageways with designations to minimize confusion
- Ensuring that every building has at least two means of egress in case one becomes blocked due to fire or smoke and that exits are remote from each other
- Ensuring that any building under construction or repair remains unoccupied until all exit facilities for the part occupied are completed and ready for use
- Protecting exit openings with an approved self-closing fire door
- Maintaining minimum width of exits of 28 inches and ceiling heights of no less than 7 feet, 6 inches
- Ensuring that public places or streets to which exits discharge are adequate in size to accommodate persons leaving the building
- Providing means of egress that are continuously maintained so that they are free of obstructions or impediments to full, instant use in the case of fire or other emergency
- Installing alarms for improper use of an exit so that the alarms cannot, even in the case of failure, impede or prevent emergency use of the exit

Employers are required to review emergency plans with employees when the plan is developed and any time the plan is changed. OSHA also recommends the following nonmandatory guidelines:

- Floor plans or workplace maps should clearly show the emergency escape routes included in the emergency action plan.
- Employees should be trained in the type of evacuation (fire, tornado, bomb threat) needed and their roles in carrying out the plan.
- Designation of refuge or safe areas should be clearly identified in the plan.
- Employees should be trained to move away from the exit discharge doors of the building and to avoid congregating close to the building, where they may hamper emergency operations and endanger their own safety.
- An adequate number of employees should be trained and available at all times during working hours to act as evacuation wardens so that all other employees can swiftly move from dangerous locations to safe areas. The agency recommends one evacuation warden for every 20 employees. Wardens should check rooms and other enclosed spaces for employees who may be trapped or otherwise unable to evacuate the area.

Many insurance carriers and loss control organizations have developed emergency evacuation plans that build on the basic OSHA regulations. Careful reading of those plans and converting the principles into practice drills can prepare employees for an emergency.

Fire prevention plans

Protecting employees from fire is more than just being sure fire extinguishers are available in the workplace. Many OSHA standards require a written fire prevention plan that lists major workplace fire hazards, proper handling and storage procedures, potential ignition sources and control procedures, and the type of fire protection equipment or systems that can control these fires.

In addition, the standards require employers to list the names and job titles of designated personnel responsible for maintenance of the equipment and systems and those in charge of fuel source hazards. Employers also must establish housekeeping procedures to control accumulations of flammable materials and residues and provide employee training in the awareness of the fire hazards of the materials and processes to which employees are exposed. Employers also must review applicable parts of the plan with each employee upon initial assignment.

Facility alarm systems are required to emit a distinct and recognizable signal. The system must be maintained by trained personnel and be tested at least annually. Backup alarm systems and lighting must be part of the safety system to take over in the event of a power failure. A public address system, intercom system, or loud hailers are recommended as additional means of communicating emergency procedures to building occupants.

The National Fire Protection Association (NFPA) is another good source of detailed information about fire safety. Guidelines and standards published by the NFPA are the basis of many communities' local fire department policies, procedures, and local ordinances.

Workplace violence

Workplace violence is a broad term applied to incidents ranging from violent thefts at retail stores to employees assaulting each other in the workplace. It covers disgruntled employees attacking their bosses as well as domestic quarrels that overflow into the workplace and involve domestic batterers stalking their victims at work. Violence at the job site is ranked among the leading causes of workplace injury and death. Millions of U.S. workers are victims of some form of bullying, harassment, or violence in the workplace each year, and about half a million nonfatal violent crimes are committed in the workplace each year. In 2010 there were 4,547 workplace fatalities in the United States, 17 percent of which were homicides.

Danger signals of potential violence

Be alert to these signs that an employee could someday turn violent:

- History of alcohol or substance abuse problems
- Known to possess or collect weapons
- Frequent displays of anger, abusive language, or threats of violence against co-workers or managers
- Other extreme behavior such as talking to self, isolation, or avoidance of social contact at work

No organization is immune from workplace violence. It occurs in law firms, large office buildings, medical offices, hospitals, classrooms, and school corridors. Without top management's support of a harassment and violence control plan, an organization can encounter tragedy at work, followed by liability claims as well as citations and fines under the General Duty clause of the Occupational Safety and Health Act of 1970. Survivors are now seeking jury trials, and company officials face the prospect of being sued as individuals for "allowing" incidents of violence to occur in their workplaces.

An organization needs clear workplace behavior rules and policies that are uniformly enforced by managers who are trained to recognize signs of potential violence. Managers should be prepared for the possibility of violence at the job site by having a formal, written plan as part of the emergency preparedness procedure. These plans and reporting procedures need to be shared with employees and supervisors.

Conduct an audit to assess the risk of workplace violence. Consider including the following elements in the audit:

- Assess both business and facility vulnerability. How safe is the building, whether owned or rented?
- Do employees deal with the public?
- Do employees handle money?
- Do employees have access to employer-owned vehicles?
- Do employees work late-night or early-morning hours?
- Do employees work alone or in small numbers?
- Are exterior parts of the property well lit and protected from intruders?
- Is accessibility to the premises controlled and are visitors escorted?
- Are personal valuables kept in locked cabinets or locked drawers?

The audit of violence potential raises awareness among employees and managers. Managers also should realize that workplace violence is increasing.

To prepare for such acts, employers should develop a plan to handle a violent incident. Having a plan and practicing it are critical to minimizing the impact of workplace violence. Establish clear rules and policies for behavior at work that deal with harassment, disruptive behavior, or the threat of violence.

An emergency preparedness committee at the management level also can take the responsibility of a violence control function, as many of the control activities are similar. Develop a formal investigative process, and communicate to employees exactly how their complaints of harassment or threats of violence will be investigated and handled.

Awareness training should be provided to managers, many of whom feel awkward about dealing with incidents of harassment and violence but don't want to give their superiors the impression that they can't manage people. Lower-level managers may need assistance from higher levels in the company or outside professional support in dealing with these incidents.

Employers also should provide alternate reporting procedures for complaints about harassment or violence. Remember, the immediate supervisor could be the harasser.

Employees should feel safe at work. For this reason, employers should promise or even guarantee that employees who report suspected wrongdoing in good faith will be protected from retaliation by any member of the organization.

Further, employers should widely publicize within the organization the procedures to air and resolve employee complaints and conflicts. Make sure workers know whom to alert if they have been threatened. Investigate promptly and thoroughly. Take prompt action to correct wrongdoing. Communicate the outcome of the investigation promptly to the employees who raised complaints.

A good first step to prevent workplace violence is to know whom you are hiring. In the recruiting and selection process, invest in a professional background check as part of a thorough applicant screening process. A background check can provide credit, motor vehicle license, and past job information for applicants.

Environmental design can help protect employees working in public contact jobs. Good lighting, good visibility to the outside, height and depth of counters, the use of bullet-resistant glass, and warning lights are all proactive steps that can be taken to minimize the risk of violence.

Finally, employers should develop a plan for responding to the media, employee families, customers, and the community during and after an incident. Appoint and train a spokesperson to speak to the media.

Don't be concerned about alarming employees when you tell them you want to conduct a vulnerability audit or develop violence control procedures. Employees are usually relieved that somebody is taking proactive measures.

Employers also should seek local police department input in developing prevention plans. Not every act of violence in the workplace can be prevented, but having a plan in place offers a good chance of preventing violence.

Alcohol and drug prevention

Federal law permits employers to test job applicants for indications of substance abuse. Such testing enables employers to develop a drug-free workplace and reduce the possibility of drug-related, unsafe work practices. Federal law and some state laws allow employers to randomly test workers in safety-sensitive jobs for drug and alcohol abuse.

How big is the problem? According to the National Institute on Drug Abuse (NIDA), an estimated 22.5 million Americans use at least one illicit drug or abuse a psychotherapeutic medication each month. About 18.1 million use marijuana monthly.

Most drug users are actively employed, raising the need for vigilance and a workplace substance abuse policy. In addition to policies, greater management awareness of this problem is needed, as is supervisor and employee awareness training.

Further, it is estimated that 88,000 deaths each year are attributable to excessive alcohol use. The economic costs of excessive alcohol consumption in 2006 were estimated at $223.5 billion. The National Institute of Mental Health estimates that nearly half of all alcoholics also have an overlapping psychological condition or problem with drug use.

Alcoholism is considered a disability under the Americans with Disabilities Act, and for that reason, alcoholic workers cannot be denied employment or terminated. However, employers can hold the alcoholic employee to the same qualifications and job performance standards as other employees—even if unsatisfactory performance is caused by the alcoholism or if the employee fails to report regularly for work or is in unfit condition to work. Although federal law does not protect alcoholics who cannot perform their job duties, some state laws require employers to reasonably accommodate the disability.

Existing company procedures can be the starting point for a formal policy on substance abuse in the workplace. Long-standing work rules prohibiting the use of alcohol and drugs on the job should be reaffirmed. Communicate management's commitment to supporting efforts to assist troubled employees. Make available the services of employee assistance programs.

The employer needs to be able to tell an employee with a drug or alcohol abuse problem that leaves of absence and treatment facilities are available. Employee assistance programs are perhaps the best known of the formal innovations to help with alcohol or drug problems and usually can arrange for drug and alcohol treatment.

Employee concerns about safety must be addressed because they may fear working alongside colleagues who have drug or alcohol problems. Supervisors need to be taught the job-related indicators of drug or alcohol abuse so that they can refer affected workers early.

Supervisors should watch for erratic work activities and ups and downs in performance, productivity, and quality. Supervisors should not make diagnoses but should refer the employee for counseling and treatment.

Employees with drug or alcohol problems are more likely to accept workplace help than home or family help. When a boss says that the job is at stake unless the employee undergoes treatment, the employee will be more likely to listen than he or she would be to listen to concerned family members.

Establish drug-testing policies. Some organizations are not comfortable with this step because of potential legal implications, but a carefully developed program can be successful.

Establishing company policies on alcohol use (for example, refusing to reimburse drinking costs on expense accounts) also can communicate messages about acceptable behavior beyond the workplace. Clear policies restricting the use of alcohol during lunchtime, in company entertaining, and at company functions convey the organization's culture. Quality of work/life programs emphasizing employee interdependence, health promotion, and employee wellness also have been shown to have a positive effect on alcohol abuse.

Incident prevention does not stand alone. Other ongoing activities, programs, and procedures in the business enterprise are preventive in nature and can help reduce injuries or minimize the extent of injuries.

Coming next ...

The next chapter outlines the importance of OSHA's role in the workplace and the impact of OSHA regulations and standards.

13

WORKING WITH OSHA

Few federal safety standards for the workplace existed in the United States until 1970, when a bipartisan U.S. Congress passed the Occupational Safety and Health Act. President Nixon then signed the act into law "... to assure so far as possible every working man and woman in the Nation safe and healthful working conditions and to preserve our human resources."

The Occupational Safety and Health Administration was created within the federal Department of Labor to administer the act. Today, complying with applicable OSHA safety standards and regulations is a requirement of doing business in the United States. Employers are responsible for knowing and understanding applicable standards. Failure to comply with OSHA standards can result in citations, significant fines, and unfavorable publicity for the employer.

OSHA is charged with the following functions:

- Developing and enforcing mandatory job safety and health standards
- Encouraging employers and employees to reduce workplace hazards and to implement new or improved existing safety and health activities
- Maintaining a recordkeeping system to monitor job-related injuries and illnesses
- Establishing responsibilities and rights for employers and employees for the achievement of better safety and health conditions
- Providing for the development of state-level occupational safety and health programs
- Providing for occupational safety and health research
- Establishing training programs to increase the number and competence of occupational safety and health personnel

The act extends to all employers and their employees in the 50 states, the District of Columbia, Puerto Rico, and all other U.S. territories under federal government jurisdiction. Coverage is provided through either federal OSHA or an OSHA-approved state plan.

OSHA's broad coverage applies to any "person engaged in a business affecting commerce who has employees." However, three groups of workers are not covered by the act:

- The self-employed
- Family members of farm employers with fewer than 10 employees
- Employees whose working conditions are regulated by other federal agencies such as the Mine Safety and Health Administration, the Nuclear Regulatory Commission, or the Federal Aviation Administration. If other federal agencies regulate some health and working conditions in a particular industry, OSHA standards apply to those areas the other agency does not cover.

Employee rights under the OSH Act

Congress designed the OSH Act to provide employee safety rights. These include the right to seek a safe and healthy workplace. Workers who believe they have been punished for exercising their workplace safety rights may contact the nearest regional OSHA office within 30 days of the time they learn of the alleged discrimination. OSHA will investigate these cases and ask the employer to restore a worker's job earnings and benefits if an employee has been illegally punished for exercising his or her rights. If necessary, OSHA can take the employer to court.

Under the OSH Act, an employee also has the right to:

- Review copies of appropriate OSHA standards, rules, regulations, and requirements that the employer is required to have available at the workplace
- Request information from the employer on hazards in the work area, on precautions that may be taken, and on procedures to be followed if an employee is injured or is exposed to toxic substances
- Have access to his or her own relevant exposure records
- Request OSHA to investigate if the employee believes hazardous conditions or violations of standards exist in the workplace
- Have his or her name withheld from the employer, be kept advised of OSHA actions concerning the complaint, and request an informal review of any decision not to inspect or issue a citation
- Have an authorized employee representative accompany the OSHA compliance officer during an inspection tour

- Observe any monitoring or measuring of hazardous materials and view these records, as specified in the OSH Act
- Review the OSHA 300 Log at a reasonable time and in a reasonable manner
- Request a closing discussion with the compliance officer following an inspection
- Object to the abatement period set in the citation issued to the employer by writing to the OSHA area director within 15 working days of the issuance of the citation
- Submit a written request to NIOSH for information on whether a substance in the workplace has potentially toxic effects in the concentrations being used and request that the employee's name be withheld from the employer
- File a complaint if punished for exercising employee rights or for refusing to work when faced with imminent danger of death or serious injury and there is insufficient time for an OSHA inspection

Section 11(c) of the OSH Act provides whistleblower protection from discrimination against employees who exercise their rights to contact OSHA regarding their safety. Employers are prohibited from retaliating against employees or taking any "adverse" actions including:

- Firing or laying off
- Blacklisting
- Demoting
- Denying overtime or promotion
- Disciplining
- Denying benefits
- Failing to hire or rehire
- Intimidating
- Making threats
- Reducing pay or hours

Employee responsibilities under the OSH Act

Employees also have responsibilities under the act. While OSHA does not cite employees for violations of their responsibilities, each employee must "comply with all occupational safety and health standards and all rules, regulations and orders issued under the Act" that are applicable.

More specifically, OSHA requires employees to:

- Read the OSHA poster at the job site
- Comply with applicable OSHA standards

- Follow all lawful employer safety and health rules and regulations and wear or use prescribed protective equipment while working
- Report hazardous conditions to supervisors
- Report any job-related injury or illness to the employer and promptly seek treatment
- Cooperate with the OSHA compliance officer conducting an inspection if he or she inquires about safety and health conditions in the workplace
- Exercise employee rights under the Occupational Safety and Health Act in a responsible manner

Employees who break OSHA regulations or facility safety rules are subject to disciplinary action by the employer.

Employer responsibilities under the OSH Act

Employers are required to inform all employees about OSHA; make copies of applicable mandatory OSHA standards available; establish or update safety and health operating procedures so that employees follow safety and health requirements; and provide training required by OSHA standards such as the Hazard Communication standard, the Hearing Conservation standard, the Bloodborne Pathogen standard, and the Lead standard.

Further, employers are required to inspect workplace conditions to make sure that they conform to applicable standards as well as minimize or reduce hazards. Employers are required to ensure that employees have—and use—safe, well-maintained tools, equipment, and personal protective equipment. Employers must use color codes, posters, labels, or signs when needed to warn employees of potential hazards.

Under the General Duty clause of the OSH Act section 5(a), each employer shall furnish a place of employment "free from recognized hazards that are causing or are likely to cause death or serious physical harm." In the absence of definitive standards or workplace safety regulations, the General Duty clause may be cited to impose a violation or penalty on any employer that does not satisfy this basic worker right.

Record keeping

OSHA also requires employers to report, record, and post information about workplace injuries and deaths. Employers must report to the nearest OSHA office within eight hours any fatal incident as well as any incident resulting in the hospitalization of three or more employees. Employers also must keep and post records of the more serious occupational injuries and illnesses.

OSHA defines an occupational injury as a cut, fracture, sprain, or amputation resulting from a work-related incident or exposure involving a single incident in the work environment. All occupational injuries must be recorded if they result in death, one or more lost workdays, restricted work or motion, loss of consciousness, transfer to another job, or medical treatment other than first aid. Each recordable occupational injury and illness must be logged on OSHA Form 301, Injury and Illness Incident Report, within seven calendar days after the initial event. This same injury must also be added to the OSHA 300 Log that is used to keep track of all the injuries in the company.

A Summary of Work-Related Injuries is calculated on OSHA Form 300A. This form records the total number of recordable cases for the company and the total number of days away from work or job transfer. Employers are required to post the OSHA 300A summary page from February 1 to April 30 of the year following the year covered by the form. Recordkeeping forms are maintained on a calendar-year basis and must be kept at the establishment for five years and made available for OSHA inspection.

Inspection protocol

To enforce safety standards, OSHA compliance officers conduct workplace inspections. These officers are allowed to enter any workplace at "reasonable times" to inspect and investigate all structures, equipment, and materials and to question privately any employer, owner, operator, agent, or employee. Workplace privacy is protected under the Fourth Amendment of the Constitution, and properties are protected from illegal search and seizure. If a business does not want a government agent to enter the premises, a warrant would first be required under Fourth Amendment protections.

OSHA uses a set of inspection priorities—attending to the worst situations first. The five degrees of priority are:

- Imminent danger situations, which can reasonably be expected to cause immediate death or serious physical harm
- Fatality or catastrophic situations, which include incidents of death or the hospitalization of three or more employees

On-site OSHA inspection procedures

OSHA has a well-defined approach and routine to workplace inspections. Before going on site, the compliance officer will review files on the nature of the business and the particular OSHA standards likely to apply.

Presentation of credentials. Most OSHA inspections begin with an OSHA compliance officer announcing his or her presence to the employer, presenting official credentials, and asking to meet with the appropriate employer representative.

Opening conference. In the opening conference, the compliance officer explains the purpose of the visit, the scope of the inspection, and the OSHA standards that apply. The employer is asked to select an employee representative to accompany the compliance officer during the inspection.

Inspection tour. The compliance officer and accompanying representatives proceed through the establishment, inspecting work areas for compliance with OSHA standards. The route and duration of the inspection are determined by the compliance officer based on the purpose of the visit: specific employee complaint, full inspection, or follow-up inspection. The compliance officer observes conditions, takes notes, may take photographs or videotape for record purposes, examines records, or takes instrument readings. OSHA inspectors are under instructions that, while talking with employees, they should minimize any work interruptions.

Compliance officers may ask to speak to employees and typically will ask the following questions:

- "Does this facility have a safety committee?"
- "Are chemicals in use here? If so, where are the Safety Data Sheets kept?"
- "Do you think your employer is interested in safety?"
- "Does the company provide personal protective equipment for use on the job?"

Closing conference. After the inspection tour is complete, the compliance officer holds a closing conference with the employer and the employee representative. During this conference, the compliance officer usually points out the unsafe and unhealthy conditions observed during the inspection and indicates all apparent violations for which a citation can be issued or recommended. The compliance officer does not indicate any proposed penalties—the OSHA area director issues penalties after receiving the compliance officer's full inspection report. The employer is informed of appeal rights once citations have been issued.

Company management should use the time between the inspector's visit and the receipt of citations to correct conditions. With documentation of corrective actions, the area OSHA director will be more likely to support a request to reduce proposed penalties.

- Employee complaints of alleged violations of standards or of unsafe and unhealthy working conditions
- Programmed inspections of high-hazard industries
- Follow-up and monitoring inspections to determine if corrections were made

Managing an OSHA inspection

Often the first contact an employer has with OSHA is when an OSHA representative appears on site. The OSHA representative may be there to perform a regular, programmed inspection or to investigate a safety complaint from a present or former employee.

The best way to deal with an OSHA inspection is to be prepared. Create a management team that is familiar with OSHA regulations and that will know what to expect and how to respond to questions.

Many organizations have found it worthwhile to develop a written policy and procedure for managing an OSHA inspection. This written policy should be disseminated to managers and supervisors at all levels.

OSHA violations and penalties

Violations of OSHA standards and rules most often result in penalties and fines. OSHA defines violations in one of six ways:

- **Willful violations,** in which the employer either knows that actions constitute a violation or is aware that a hazardous condition exists and makes no reasonable effort to eliminate it. Willful violations are the most serious types of OSHA violations, and penalties can reach as high as $70,000. A willful violation resulting in the death of an employee is punishable by a court-imposed fine, imprisonment of up to six months, or both.
- **Serious violations,** whereby a substantial probability exists that a hazard could result in death or serious physical harm and the employer knows, or should know, about the hazard. A mandatory penalty of up to $7,000 for each violation is proposed.
- **Other-than-serious violations,** which have a direct relationship to job safety and health but probably would not cause death or serious physical harm. A penalty of up to $7,000 is discretionary with the OSHA area director.
- **Repeat violations** can result in fines of $7,000 for each day the violation continues.

- **Failure to correct prior violations** may bring a civil penalty of up to $7,000 for each day the violation continues beyond the prescribed abatement date.
- ***De minimus* violations** have no direct impact on job safety and health, such as not keeping required records, and typically carry no penalty.

OSHA has the responsibility to assess penalties based on the gravity of the violation, the size of the business, the good-faith efforts of the employer to abate the cited violations, and the employer's history of previous safety and health violations. A penalty must be proposed for serious violations, but employers with strong hazard detection and control activities in place may have fines reduced by up to 50 percent. Employers with no such activities in place cannot qualify for reductions.

Smaller organizations frequently get a break under OSHA regulations. For example, a company with fewer than 25 employees may qualify for a 60 percent reduction in fines, while organizations with up to 250 employees may receive reductions from 20 to 40 percent.

OSHA assistance

OSHA encourages and assists employers in developing proactive safety and health activities. Employers may request consultation assistance, under the condition that any hazards identified by the consultant must be corrected. Comprehensive consultation assistance includes an appraisal of mechanical, physical, work practice, and environmental hazards of the workplace and of the employer's present job safety and health activities.

Employers who receive a comprehensive consultation visit, correct all identified hazards, and demonstrate that an effective safety and health plan is in operation may be exempted from OSHA programmed inspections (although not from complaint or incident investigations) for a period of one year.

Another part of OSHA's effort to extend worker protection is its Voluntary Protection Program (VPP)—a partnership with employers and workers designed to expand worker protection beyond the minimum goals required by OSHA standards.

Hundreds of worksites employing hundreds of thousands of workers participate in this model program. According to OSHA, employers who meet VPP eligibility criteria typically experience lost workday and injury incident rates 50 percent lower than the overall industry averages, reduced workers' compensation costs, increased productivity, better morale, and reduced absenteeism.

OSHA's three Voluntary Protection Programs—Star, Merit, and Demonstration—are designed to recognize the outstanding achievements of companies that have successfully incorporated comprehensive safety and health programs into their total management systems and to establish a relationship among employers, employees, and OSHA that is based on cooperation rather than coercion.

The Star program may apply to any employer in any industry who has successfully used a comprehensive safety and health process to reduce injury rates below the national average for the industry. Specific requirements for the program include management commitment and employee participation, a high-quality worksite analysis program, hazard prevention and control programs, and comprehensive safety and health training for employees. All these requirements must be in place and operating effectively.

The Merit program is a stepping-stone to Star program participation. An employer with a basic safety and health program built around the Star requirements that is committed to improving the company's safety and health activities and that has the resources to do so within a specified period of time may work with OSHA to meet Star qualifications.

The Demonstration program is for organizations providing Star-quality worker protection in industries where Star requirements may not be appropriate or effective. It allows OSHA to recognize outstanding safety and health activities in its own way.

OSHA standards

Many OSHA standards are consensus standards recommended by management-supported professional and technical organizations such as the National Fire Protection Association. Federal regulations are written by a federal agency like OSHA and are published in both the *Federal Register* and the Code of Federal Regulations. They are the legal means of implementing and enforcing any law, act, or executive order.

Two different types of standards are used to cover a variety of working conditions and locations. A Horizontal Standard is one that covers a wide range of general industries and activities. Hearing conservation and protection, for example, are required for any industry that produces hazardous levels of noise. This could occur in a range of different industries, from construction to manufacturing.

The other type of standards are called Vertical Standards. These standards tend to apply specifically to one industry or segment of the workforce. An example of these standards would be those specifically written for the construction industry or agricultural workplaces.

The following are some OSHA standards affecting general industry.

- **Access to Employee Exposure and Medical Records** gives employees, OSHA, and union representatives the right to review employee medical records and workplace health exposures that are maintained by the employer.
- **Air Contaminants and Other Chemicals** regulates exposure limits to air contaminants.
- **Asbestos** outlines permissible exposure limits, respiratory protection, warning signs, and protective clothing for persons working with asbestos and similar products.
- **Benzene** deals with occupational exposures to benzene, excluding motor fuels, gasoline, or other fuels subsequent to its discharge from bulk wholesale storage facilities.
- **Bloodborne Pathogens** requires employers of health care, medical, dental, and first aid workers to protect employees involved in treatment and to cover the proper prevention measures to protect these employees from infection.
- **Cadmium** applies to all occupational exposures to cadmium and cadmium compounds in all industries covered by OSHA.
- **Control of Hazardous Energy** requires affixing lockout or tagout devices to prevent unexpected start-up of equipment and requires inspections and employee training.
- **Electrical** outlines federal design safety standards for electrical systems, work practices with electricity, maintenance, and related requirements.
- **Emergency and Fire Prevention Plans** applies to all emergency action plans required by a particular OSHA standard.
- **Emergency Flushing** requires special facilities for the quick drenching and flushing of eyes and body when exposed to corrosive materials.
- **Fire Protection** outlines requirements for employer-established fire brigades and fire extinguishing systems.
- **Formaldehyde** deals with occupational exposures to formaldehyde and materials releasing formaldehyde.
- **Hand and Portable Powered Tools** requires basic guarding mechanisms for hand and powered tools found effective in protecting employees using this equipment.

- **Hazard Communication** requires evaluation of chemical hazards in the workplace and the communication of those hazards to employees through labels, Safety Data Sheets, training, and a written local hazard communication program.
- **Hazardous Materials** deals with handling designated hazardous materials, piping systems, and outside container storage.
- **Laboratory Chemicals** applies to all employees engaged in the laboratory use of hazardous chemicals, as defined by OSHA.
- **Lead** outlines permissible exposure limits for lead, provisions for medical examinations, and abatement procedures.
- **Machinery and Machine Guarding** requirements cover machinery, including abrasive wheels, mills and calendars, mechanical power presses, forging machines, and mechanical power transmission apparatus.
- **Materials Handling and Storage** includes OSHA regulations on powered industrial trucks, cranes, derricks, and slings.
- **Means of Egress** (exiting from a facility) requires continuous and unobstructed exit from a building or structure.
- **Medical and First Aid** provides for worksite first aid and medical coverage.
- **Occupational Noise Exposure** requires effective engineering or administrative controls to reduce workplace noise levels equal to or exceeding 85 decibels on an eight-hour time-weighted average.
- **Personal Protective Equipment** requires eye and face protection, respirators, and head and foot protection.
- **Process Safety Management of Highly Hazardous Chemicals** requires preventing or minimizing the consequences of the release of highly hazardous chemicals.
- **Special Industries** includes precautions for specific industries including textiles, bakery equipment, laundry machinery, sawmills, pulpwood logging, and telecommunications.
- **Ventilation** provides for safe ventilation of a multitude of materials used in industry and construction.
- **Vinyl Chloride** applies to the manufacture, reaction, packaging, storage, handling, use, or transportation of vinyl chloride or polyvinyl chloride and employee exposure to these chemicals.
- **Walking/Working Surfaces** covers dangers involved with slips, trips, and falls on walking and working surfaces.
- **Welding, Cutting, and Brazing** outlines the safety precautions required while welding, including fire precautions.

Figure 13.1. Occupational Exposure to Excessive Noise May Require Hearing Protection

(Adam Gregor/iStock/Thinkstock)

Additional federal laws

Other federal laws pertain to workplace safety. These include:

- **The Americans with Disabilities Act** protects people with certain defined disabilities from discrimination in employment, public services and transportation, public accommodations, and the availability of telecommunications. A disability is defined as a mental or physical condition that substantially limits "a major life activity," as defined in the act. On request, employers are required to provide qualified individuals who have disabilities with reasonable accommodations to assist them in performing the essential functions of the job, unless the accommodation would cause undue financial hardship to the employer or a significant safety risk to the employee or co-workers.
- **The Drug Free Workplace Act** requires employers with contracts with the federal government of $100,000 or more to meet certain posting and recordkeeping requirements and to develop policies prohibiting the unlawful manufacture, distribution, possession, or use of controlled substances in the workplace.
- **The Rehabilitation Act of 1973** prohibits discrimination against people with disabilities by contractors with federal government business totaling

$10,000 or more per year. Contractors with $50,000 or more in annual government business and with 50 or more employees must prepare an affirmative action plan to comply with the act.

- **The Toxic Substances Control Act, the Federal Insecticide, Fungicide, and Rodenticide Act,** and **the Noise Control Act,** among other laws, fall under the jurisdiction of the Environmental Protection Agency, which is responsible for regulating the quality of air and water and almost all other environmental problems.

Coming next ...

The next chapter will show that occupational health is an important part of an occupational safety hazard detection and control process.

14

OCCUPATIONAL HEALTH

Occupational illnesses are nontraumatic work-related health incidents that are sometimes difficult to identify as work-related. Examples of occupational illnesses are work-related dermatitis, lung disorders, and musculoskeletal disorders that are not easily cured or corrected and that have expensive, long-term implications. Illnesses that begin as off-the-job illnesses may be intensified by work exposures, at times making hazard detection difficult to determine.

Occupational health problems

Hundreds of new chemicals are introduced into the workplace each year. Combinations of these chemicals, new materials, and new processes present an array of potential hazards for employees. Even traditional, low-hazard office settings now have ergonomic hazards from repetitive motion, eyestrain from inadequate lighting, and environmental hazards from poorly ventilated structures.

If workers report job health problems, then an employer should ask for a health hazard evaluation of the facility. Large organizations with in-house medical staff and industrial hygienists can tackle such problems quickly. Smaller organizations may have to use outside resources such as local industrial medical clinics, industrial hygiene services, or occupational dermatologists, physical therapists, or ergonomists.

Some state departments of health perform hazard evaluations without charge to employers and involve management, employees, and union representatives in their survey work and report of findings. The National Institute for Occupational Safety and Health offers the same type of service on a selective basis.

Occupational exposure sites

How can harmful substances to which workers are exposed on the job get into their bodies? A toxic substance can enter the body through four routes of exposure:

Inhalation involves breathing the airborne toxic substance into the lungs.

Absorption involves taking in a toxic substance through the skin.

Ingestion is the process of swallowing toxic materials as a result of eating in a contaminated work area, or having contaminated fingers or hands while smoking or eating.

Injection is an unintentional skin puncture by a needle or instrument contaminated with infected blood or tissue.

A number of factors in the workplace can cause sickness, impaired health, discomfort, or inefficiency. They are frequently classified as chemical, physical, biological, or ergonomic factors.

Chemical hazards can arise from excessive concentrations of mists, vapors, or gases that form dust and fumes in the air.

Physical hazards include excessive levels of heat, radiation, noise, vibration, and extremes of temperature and pressure.

Biological hazards can come from insects, molds, fungi, and bacterial contamination of food and water.

Ergonomic hazards stem from poorly designed work areas or tools, improper lifting or reaching activities, or repeated motions in an awkward position.

Symptoms of exposure to these hazards can be immediate (acute) or can appear after longer exposure at a lower intensity over time (chronic). Some types of occupational health problems (dermatitis, hearing loss, industrial asthma) may not appear immediately but take time—perhaps years—to develop into a work-associated health problem. It is always important that workers know the signs and symptoms of exposure to potential occupational hazards in order to properly identify, minimize, and control them. Just knowing the smell associated with a toxic chemical can save a life. Knowing that overexposure can lead to headaches can be equally useful.

Focusing on job illnesses

Management needs to know and understand the chemicals and other materials in their workplace. A valuable guide to exposures and the corrective steps to take to avoid overexposure are Safety Data Sheets, which should be at every worksite and readily available to any worker.

Common occupational health problems

The facility hazard evaluation process should indicate hazards or significant exposures and other, less serious hazards. The most serious hazards—those that can potentially cause serious employee health problems—obviously should receive attention first.

Employers in general industry, distribution centers, construction jobs, and office settings face many common hazards. The following sections define the extent of each of these common problems and outline some basic hazard control steps that management should consider when they encounter these hazards, adapting these suggestions to the specific hazards in their facilities.

Back strains and injuries

Back strains are considered part of the broad area of musculoskeletal disorders. Millions of workers must lift, bend, or pull as part of their everyday job duties. Jobs such as garbage collector, warehouse worker, mechanic, and nursing aide are the leading jobs in which employees can develop back problems. But sedentary workers in office jobs, at computer workstations, and at workbenches can also have problems when they sit for prolonged periods.

To gain control of back injuries and strains in the workplace, management should:

- Focus on the jobs that are causing these strains, using data from first aid logs, the OSHA 300 Log, and workers' compensation reports
- Conduct a hazard analysis, determine the exact steps or motions causing the problems, and explore redesigning the job to eliminate as many of the stresses as possible
- Use mechanical equipment to replace heavy lifting tasks wherever possible
- Provide training in safe lifting techniques
- Encourage employees to become more physically fit with regular exercise and stretching
- Use adjustable, ergonomically designed chairs for jobs that require a stationary position or sitting positions

Ergonomics problems

Cumulative trauma disorders, also known as repetitive motion injuries and musculoskeletal disorders, result from repeated use of the same part of the hand, wrist, or arm. These injuries develop as a result of repetitions of a particular motion, forceful repetitions of the motion, or awkward positioning of the hand, wrist, or arm in completing the motion.

The definition of cumulative trauma disorders includes a broad array of medical conditions such as carpal tunnel syndrome, tendonitis, tenosynovitis, tennis elbow, trigger finger, Raynaud's syndrome, and DeQuervian's disease. In a typical case of carpal tunnel syndrome, tendons swell in the wrist and put pressure on nerves, resulting in reduced motion ability and fingers that are no longer able to grasp objects firmly. Cumulative trauma disorders can result in nerve damage and disabling numbness and pain that may require surgery.

The science of ergonomics is dedicated to overcoming cumulative trauma strains by developing a better fit among worker, task, and workstation. It draws on disciplines such as biomechanics, physiology, physics, statistics, and basic engineering.

Musculoskeletal problems are a major recordable occupational illness under OSHA regulations. A management plan to overcome these cumulative trauma disorders can take advantage of some of the following ergonomic protocols:

- Zero in on ergonomic problems at the facility—the first step in a hazard elimination and control process. Some problems are self-evident (e.g., employees complain or lose time from work), but a formal review also should be made of first aid and medical records, the OSHA 300 Log, injury reports, and the employer's first report of injury to the workers' compensation carrier.
- Include early medical involvement to help identify and evaluate the first signs and symptoms of cumulative trauma disorders and control further spread of the problem. Alternate work or temporary light-duty programs can be used to help employees remain at work doing other jobs while they recuperate.
- Conduct a thorough survey of those jobs or processes causing the greatest problems. The employer should then focus on steps to eliminate or control ergonomic hazards using workstation redesign or rearrangement, ergonomically designed tools, and ergonomically safe work practices. A next step is to train employees in the ergonomically safe work practices.
- Follow up. Without continuing motivation and follow-up, workers return to old ways quickly.

Figure 14.1. Ergonomic Hazards Associated with Word Processing and Office Work *(Ryan McVay/Photodisc/ThinkStock)*

Some steps to ergonomic improvement involve simple, inexpensive common sense. Listen to employees working these jobs; many workers have simple, cost-effective ideas to simplify their work and ease the strain on their bodies. On the other hand, revisions to building and equipment will cost money and take time to implement.

Chemicals and hazardous substances

More than half a million chemical products are considered hazardous under the OSHA Hazard Communication standard. Tens of millions of U.S. workers are exposed to chemical hazards. These chemical exposures may cause or contribute to serious health effects such as heart ailments, kidney and lung damage, sterility, cancer, burns, and skin rashes.

The federal Hazard Communication standard (often referred to as the Employee Right-to-Know law) requires employers who use hazardous chemicals to do the following:

- Prepare a written inventory of those chemicals—the important starting point to knowing what chemicals are on hand. Chemical manufacturers, importers, and distributors are required to ship chemicals with a label, tag, or other method to identify the particular chemical in the container.
- Train employees about the chemicals they work with through signs, labels, and Safety Data Sheets.
- Have Safety Data Sheets available to supervisors and employees for every chemical used in the facility.
- Prepare a written program summarizing the employer's hazard communication program, updating it as elements change.

Bloodborne pathogens

Millions of workers in health care facilities and medical and dental offices are exposed to potential infection from contact with blood products. Contamination incidents can lead to exposure to hepatitis B virus or human immunodeficiency virus (HIV), which causes acquired immunodeficiency syndrome (AIDS). Principally designed to protect workers in hospitals, nursing homes, blood banks, and medical and dental offices, the federal law also applies to occupational medical departments; first aid personnel assigned by their employers to treat cuts, burns, and wounds; body-modification professionals (tattooists and piercers); and several other occupations.

The federal standard requires employers to establish a written exposure control plan and calls for the use of puncture-resistant containers for used needles, work practices such as hand-washing to reduce contamination, and the use of personal protective equipment such as gowns and gloves. A written schedule is required for cleaning, discarding contaminated instruments and needles, and handling regulated wastes; written procedures are required for handling contaminated laundry.

The standard also requires that the hepatitis B vaccine be offered at no charge to personnel who handle body fluids or treat injuries. If a serious incident should occur, the employer is required to make a medical evaluation available to the employee(s), including, if necessary, laboratory tests, postexposure preventive treatment, and any follow-up counseling that may be appropriate.

Medical information collected in the implementation of the standard is to be kept confidential, in locked medical files separate from the regular employee personnel file.

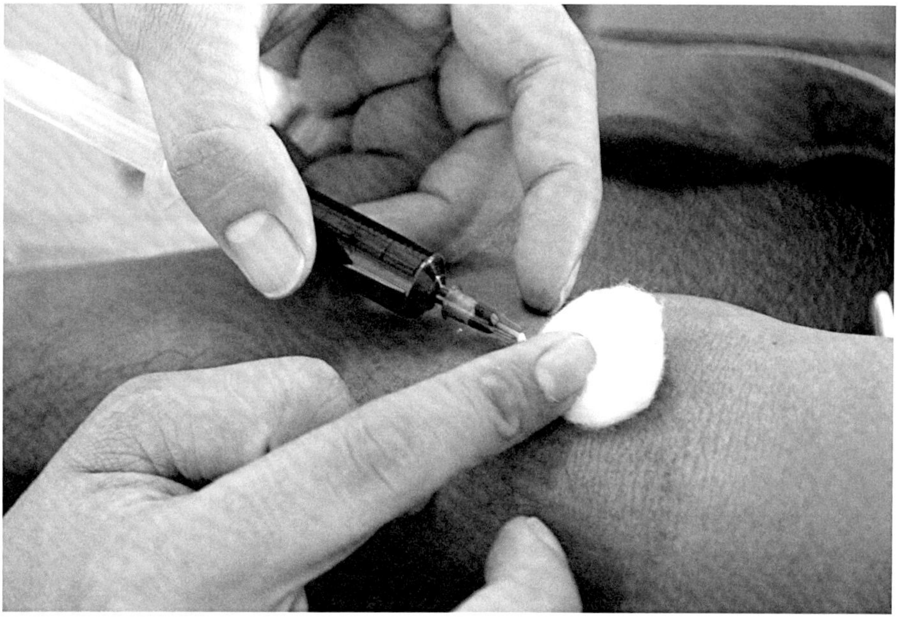

Figure 14.2. Workplace Hazards Associated with Exposure to Bloodborne Diseases *(MeePoohyaphoto/iStock/ThinkStock)*

Workplace noise

Impaired hearing is one of the most common physical workplace injuries. Approximately 30 million workers are exposed to hazardous noise on the job, and an additional 9 million are at risk for hearing loss from other agents such as solvents and metals. One out of 10 people has some degree of hearing loss. Job-related exposure to high noise levels without adequate hearing protection can have severe health implications over time.

In 2008, approximately 2 million U.S. workers were exposed to noise levels that put them at risk of hearing loss. Reported cases of hearing loss accounted for 14 percent of occupational illness in 2007.

There are three reasons why hearing loss is one of the most tragic workplace injuries. First, hazardous levels of noise are extremely easy to identify. Basically, if you have to yell to be heard, as in trying to have a conversation with someone pushing a lawn mower, you are likely at or over 85 or 90 dBA. This is at or over the safe limit of worker exposure for an eight-hour day. At this level, an employer should arrange for the monitoring of noise levels throughout the workplace and for the duration of the workday. Professionals such as industrial hygienists can conduct a noise survey with sound level meters or personal noise level recorders.

The second reason why exceeding noise limits is so unacceptable is that we are well aware that the levels of noise mentioned above, for extended durations, cause irreversible hearing damage. Chronic exposure over time causes damage, and hearing cannot be replaced.

The third reason why worker hearing loss due to excessive noise exposure is unforgivable is that it is often extremely easy to take protective action against hearing loss occurring. Typically, engineering controls such as damping or barriers can be used to reduce the noise levels. But no matter what, as a last resort, personal hearing protection in the form of earmuffs or earplugs can be very effective in reducing workers' exposure to hazardous levels of noise.

Any workers subjected to more than 85 dBA are required to be enrolled in a hearing conservation program. An effective program should include the following elements:

- Noise monitoring should be conducted and, where noise levels exceed 85 decibels on an eight-hour time-weighted average, a formal, written hearing conservation program is required, including mandated hearing protection for certain jobs. A variety of hearing protection is available, depending upon the noise levels in the facility.
- A planned program of engineering controls, administrative controls, and use of personal protective equipment should be developed.
- Newly hired employees should receive beginning or base-level audiograms at the time of hire to determine their hearing levels before starting the job. These employees should then be fitted with hearing protection suitable for the noise exposures.
- Employees exposed to noise levels of 85 decibels or higher should receive an annual audiogram and refresher training in the correct wearing and fitting of hearing protection.
- Employees should receive audiogram test results within 21 days, and if hearing loss is confirmed, the incident should be posted on the OSHA 300 Log of Injuries and Illnesses, per requirements. Workers with the onset of hearing loss should receive refresher training in the correct use of hearing protection.

Motor vehicle injuries

The largest category of fatal job-related injuries is comprised of motor vehicle crashes. Motor vehicle safety should be followed not only by drivers of over-the-road tractor trailers and buses but also by anyone who drives motorized vehicles on company property and anyone who travels on business.

Safe vehicle operation involves safe vehicles, trained and alert operators, and good driving habits. Motor vehicle safety control programs for cars or trucks should include the following elements:

- **Careful driver selection**—Employment applications should inquire into the applicants' driving records and request that applicants make available their motor vehicle records to be verified or authorize the employer to request each applicant's driving record from the appropriate state.
- **Physical examinations**—Department of Transportation regulations insist upon physical examinations and driver testing for operators of vehicles weighing more than 10,000 pounds.
- **Valid licensing**—A basic requirement of all drivers is that they have valid operator's licenses for the classes of vehicles they are operating.
- **A driver road test**—The applicant should go through a "live" road test with a company representative so that the applicant's driving performance can be observed.
- **Care in operation of equipment**—The employer should review the importance of a pretrip check of the vehicle, including checking for properly inflated tires, clean windshields, correct angles of rearview mirrors, brakes working properly, and an understanding of the vehicle's gauges and what they tell the driver. The Federal Highway Administration requires daily vehicle inspections before and after each trip and the prompt correction of any defects noticed.
- **Defensive driving techniques**—The employer should explain the use of defensive driving and courtesy as a defensive driving technique. Company policy and most state laws require the use of occupant restraints at all times when a vehicle is in operation.
- **Incident reporting**—The employer should outline the company's policy and procedures for reporting the incident and arrangements for medical treatment if a crash, driver injury, or truck or property damage occurs.
- **Performance review**—Employees' driver safety performances should be included in the job performance review programs of those who operate company vehicles.
- **Acknowledgment of achievements**—Honor safe drivers in the company and publicize their good records widely.

Indoor air quality

The Environmental Protection Agency estimates that one in three buildings in the United States is contaminated with poor indoor air quality, which affects employee health and productivity.

Common symptoms of poor indoor ventilation include headaches, dizziness, nausea, eye irritation, sneezing, coughing, sinus congestion, chest tightness, and shortness of breath.

Inattention to the maintenance of heating, ventilation, and air conditioning equipment can result in dirty air systems, clogged or inefficient filters, contaminated air ducts, and mold or bacteria growing in ventilation systems or behind walls. One of the most common sources of indoor mold is acute or chronic water intrusion into buildings and homes. Mold will begin to grow on wet building materials such as wallboard or carpeting within 24 to 48 hours. If the water source is not stopped and the materials are not dried out within this period of time, mold will take hold and begin to grow. Even if the material eventually dries out, the mold spores may remain for years. As soon as the next water intrusion event occurs, the mold will immediately begin growing again and will send out spores and mycotoxins into the air. In such cases, the only solution is to remove all of the contaminated materials and replace them with new building materials—but only after the water intrusion source has been permanently eliminated.

Today, the technology exists to create and maintain acceptable indoor air quality, and most indoor air quality problems can be solved. The following steps should be considered when addressing problems of indoor air quality:

- Review Safety Data Sheets for materials that are suspected contributors to the pollution. Oversee the work practices of contractors applying sealing compounds, wall and carpet adhesives, and paints and varnishes.
- Make sure the ventilation system is clean and free of stagnant water and damp filters. Be sure to open the ventilation system to its engineering limits and ensure that roof vents are open and unobstructed.
- Focus on air recirculation, which may be as high as 90 percent in some buildings.
- Be aware that faulty cleaning procedures, caused by improperly diluted industrial-strength cleaners, may be leaving chemical residue in the building's carpeting.
- Watch for ordinary but often overlooked hazards such as engine fumes emanating from trucks parked in a garage or next to a building. Carbon monoxide fumes rising in an elevator shaft from a basement parking area have caused problems in some buildings. Air testing may be required.
- Duplicating machines and processes using solvents should be moved to enclosed areas with good air exchange.
- When planning new facilities, examine the possibility of installing windows that can be opened to let in fresh air and carbon dioxide sensors at key points in the ventilation system.

Coming next ...

The next chapter will describe a variety of methods for continuous improvement of the safety program over time. New types of management systems that set goals and policies also are introduced.

15

CONTINUOUS IMPROVEMENT AND MANAGEMENT SYSTEMS

After experiencing modest successes in hazard detection and control, some managers, supervisors, safety committees, and workers will feel a sense of exhilaration: "We're making headway!" they'll proclaim. And that may be true. But it is not the end of hazard detection and control or of incident prevention activities.

Incident prevention is part of the safety and health management process, and team-based safety efforts in which workers are genuinely involved and empowered to take steps to advance the employer's objectives are almost always more successful than top-down management styles. It also makes the workload easier for managers and supervisors.

Safety coordinators cannot take on for workers the role of being safe, nor can employees be supervised into being safe. Instead, successful incident prevention activities involve many people, at all levels of the organization.

Ideas that work

The following ideas are characterized by some safety experts as elements of long-term safety processes that achieve hazard recognition and control objectives:

- Convince management that safety can be managed.
- Speak the language of top executives by using language that speaks to the bottom line.
- Arrange for the top managers of the organization to receive reports of every injury or near miss within 24 hours of its occurrence.
- Set up safety performance improvement goals using data from past experiences and benchmark performance in the industry.

- Establish line management accountability for safety performance.
- Teach the line workers how to conduct safety inspections, with emphasis on both unsafe conditions and unsafe acts.
- Establish a management steering committee to ensure the long-term success of the safety process.
- Applaud successes.
- Work with key management members to develop a five-year plan for safety and health management.

Safety roles

Who does what in safety? Safety coordinators who work closely with line managers usually do not have the authority to change operating rules and procedures. Safety and health coordinators should therefore act in support or consulting roles to line managers. Safety can be seen as a service function in the organization by helping supervisors to make jobs safer and more efficient at the same time.

Role of line manager in incident prevention

A line manager's safety responsibilities include the following:

- Setting an example by working safely
- Emphasizing safe work behaviors on daily facility tours
- Correcting hazards or unsafe work practices on the spot or implementing recommendations for hazard elimination and correction
- Supporting positive safety activities, such as incident investigation, safety inspections, and effective use of personal protective equipment
- Making safety a partner with quality, efficiency, and customer service

Role of safety coordinator in incident prevention

A safety coordinator's responsibilities include the following:

- Assisting line managers in carrying out their safety responsibilities
- Assisting in coordinating safety activities determined by line managers to be the most appropriate for the business and its needs
- Providing information on evolving state and federal laws on safety, occupational health, and environmental controls
- Ensuring that an injury-reporting and recordkeeping system is in place and operating well
- Acting as a resource to line managers by providing assistance, information, and materials without doing their jobs for them

Example: *As new, safer chemicals become available, it may be possible to exchange a toxic chemical used in a process for a safer one. Line supervisors may not have the time or resources on hand to investigate or test new alternatives as they become available. However, a safety coordinator can obtain information on new chemicals, analyze their hazardous properties, find cases where the chemicals are being used effectively at other locations, and help implement testing and application at the given facility.*

Recognition and rewards for safety success

Many people view on-the-job safety as a basic personal responsibility and consider awards or rewards for self-protection inappropriate and self-defeating. They ask, "Should employees receive extra rewards for just coming to work and working safely?"

Others take a broader view, believing that successful hazard detection and control activities should be highlighted and applauded and that those employees who contribute to such activities should be acknowledged in some way. Recognition programs are a proven method of enhancing performance, increasing morale, and building employee loyalty.

Safety recognition incentive programs have been around a long time. Earlier versions of safety incentive programs provided rewards for "incident-free days" or reduction in numbers of injuries. Sometimes these programs gave rewards through a lottery or safety bingo or included across-the-board rewards for all workers.

Such approaches to safety incentives are still used, but concern has increased that safety incentive programs rewarding employees for low incident rates send the wrong message. Some critics, particularly labor groups, are concerned that low incident rates can be achieved by low reporting of injuries, by hiding injuries, or by not having injuries treated.

The real question is whether rewards such as bonuses improve safety or just suppress reporting. If bonuses are based on safe behavior, they can improve safety. But if they are based on "not having an incident for a period of time," workers may be tempted to not report injuries.

Some basics should be considered before starting any safety recognition or reward program. An awards program should never be a substitute for an effective safety process, which must be in place and operating well before any form of safety award is introduced.

Types of safety recognition approaches

Two types of safety incentive programs exist: traditional awards rewarding reductions in the number of injuries and illnesses reported and the more positive type of rewards for safe behaviors resulting in fewer injuries.

Keep the following points in mind when establishing any safety incentive program:

- A successful safety activity must come first. When installing a new or revised safety process, allow time for the process to become embedded in the organization's culture before announcing an award framework. This integration may take at least a year or two.
- Write a statement for the rewards or recognition activity that ties in with the organization's business and safety goals.
- If a traditional awards program is being considered, think through the benchmarks: X number of days since the last lost workday case, percentage reduction in first aid cases, or percentage reduction in workers' compensation costs. But make sure that this approach to awards does not result in workers hiding injuries. Determine the time duration for the incentive program and specify start and end dates.

With safety incentive programs, the danger exists that employees will not report injuries or first aid cases because of the fear of co-worker and management resentment. Therefore, any cases uncovered after the fact should be debited to the program and widely publicized. Alternately, there should be no penalty if the employee or groups of employees do not reach the stated goal.

- For behavior recognition awards, consider honoring all who served on the safety committee during the past year, those who acted as safety inspectors, the incident investigation team judged to have provided the best recommendations to prevent recurrence of an incident, or the person who entered the best incident reduction or safety suggestion of the year. People who have avoided serious injury through the wearing of personal protective equipment also may be honored.
- Develop the baseline of information to measure achievements of the incentive program, solicit input from line and staff members, and then publicize widely the statistical starting point of your program.
- Set goals that are reasonable and attainable. Develop a written rationale for selecting those goals; management will want to rely on that rationale in the event of any protests regarding the awarding of prizes.
- Develop a budget for the awards based on some percentage of anticipated savings. This will dictate whether awards are inexpensive, such as T-shirts for the winners, or are more expensive items, such as a $50 savings bond for each employee.
- Solicit input from line and staff managers and the safety committee. Modify the program accordingly.

- Carefully think through what is being chosen as a recognition award. Decide whether the awards are to be for individuals or for groups. What might work for plant employees may not be attractive to sales personnel.
- Advertise and promote the program. As much value will come from a strong communications effort and employee enthusiasm in support of safety as from the awards themselves. Provide regular communication updates on progress toward goals.
- Award prizes promptly. As the end of the award period approaches, be ready so that announcements of the good news can be made within hours of the end of the period. Award the prizes promptly. Publicize why the incentive is being awarded.

Awards should be simple, attractive, and inexpensive items. Awards programs should never replace safety training, safety meetings, or supervisory leadership of safety.

Awards and rewards for supervisors

Good hazard detection and control performance should be acknowledged prominently, and managers, supervisors, and employees who are active in such positive activities should be rewarded. Positive reinforcement makes for the continuation of good work.

If supervisors are, in fact, leaders of the hazard detection and control activity, then special awards may be in order for them. They may share in worker awards or be provided with an above-average merit increase or a one-time bonus honoring good performance. Gaining favor these days are alternate monetary rewards: year-end incentive payments or even one-time awards of shares of company stock. These have a direct effect on supervisory compensation and make a positive impression.

Special honors for a supervisor at a company function can work wonders for morale. Some organizations present plaques or other types of visible awards for achievements of supervisors.

Whatever recognition method is used can enhance the work of supervisors, who are the leaders of injury prevention and control activities.

A management systems approach to incident prevention

An organization's response to incident prevention depends largely on its safety culture, which reveals itself through the company's values and norms.

What are the most important business objectives of the organization? Perhaps the values in an organization are to be a leader in production, customer service, and quality performance. But there may be others. Safety does not need to be first, but it should be equal with these other objectives.

A management systems approach to incident prevention can have the same elements that are used to achieve important business objectives. The approach should include the following elements:

- Clear expectations for occupational safety and health established by senior management
- Significant management and employee involvement
- Establishment of action plans to achieve these objectives
- Safe work practices based on job safety analyses
- Training of workers at all levels in the safe work practices
- Providing safety and health resource staff with the necessary training to be highly knowledgeable in technical and regulatory requirements
- Observing work practices and providing feedback on positive practices and correction of negative ones
- Tracking of performance results against the objectives

If productivity is a key business objective, safety can positively impact productivity, whereas injuries and illnesses negatively impact productivity.

Safety performance monitoring and measurement

Safety performance evaluation enables continuous monitoring and regular assessment of the safety levels achieved by an organization. Safety performance should be verified in reference to the organization's safety policy, safety objectives, and safety performance targets using a dedicated set of safety performance indicators. In order for performance indicators to be useful, they must be:

- Specific
- Measurable
- Achievable
- Realistic
- Traceable over time

Effective use of the performance indicators allows trends to be analyzed in order to identify when actions are necessary to correct observed deterioration of the safety program.

Information for safety performance monitoring comes from a variety of sources, such as:

- Surveillance, audits, reviews
- Surveys
- Occurrence reporting
- Investigation of accidents, incidents, and near misses
- Safety studies
- Systematic capture of data reflecting actual day-to-day performance
- Continuous monitoring of day-to-day activities related to operational activities

Each of these types of information sources may be available to an organization. However, individual organizations will need to tailor the sources to the appropriate scope and scale.

The size and structure of the organization and the operational environment have to be considered when setting up safety monitoring arrangements in the organization. In general, the following steps should be taken in the collection and analysis of data:

- ***Identification of indicators to be monitored***—technical and organizational, quantitative or qualitative, leading (proactive) or lagging (reactive)
- ***Collation of the information***—systematic collation and evaluation of results from all safety monitoring activities to ensure that interrelationships can be detected
- ***Analysis of indicators***—analysis of indicators and establishment of trends and related causes and influencing factors
- ***Application of corrective action process***—corrective actions being determined, taken, and followed up wherever the monitoring shows that an element is approaching a point where safety may be affected to an unacceptable extent; coordination with relevant units/organizations taking place wherever necessary; the indicators and their evolution being documented as well as actions and their results being taken

Success may be a series of small steps

Progress in incident reduction for the safety coordinator may be a series of small steps: converting the stubborn supervisor, who eventually begins to realize the value of incident prevention; achieving higher priority of company funds for safety items; or realizing that injury prevention discussions are now a main agenda item with operating managers.

For the safety coordinator, some personal attributes will be key to achieving those steps to success. They include the following:

- **Being patient**—Safety coordinators shouldn't expect to see a 100 percent transformation toward incident prevention in the organization in the first few months on the job.
- **Being an influence**—A leader is a person who influences change, and a safety coordinator usually is chosen because he or she is persistent and, in a friendly manner, tries to influence others in the right direction. Some people will take months to be converted. However, the higher in the organization the influence reaches, the greater effect the influence will have on others in the organization.
- **Working with others**—The safety coordinator cannot do the safety job alone. The real responsibility of a safety coordinator is to get line managers to do their jobs of hazard detection and control. That in turn will make incident prevention as important as the prevention of quality defects, cost inefficiencies, and production snafus.
- **Using personal qualities**—Busy managers who have many other things on their minds will believe in a sincere person.
- **Developing professionally**—The typical safety coordinator starts off his or her assignment with little or no background in occupational safety and health. Influence will grow in direct measure to his or her knowledge and experiences accumulated from dealing with occupational safety and health.
- **Helping others focus on safety**—Managers may have other things on their minds—other distractions, other priorities. But the safety coordinator can help managers focus on the value of incident prevention.

A matter of business excellence

When operational discipline becomes an integral part of the work environment, programs on paper become programs in action. The process of establishing operational discipline and operational excellence requires time and hard work. But the result is business excellence. Successful safety processes involve learning from mistakes or, in the management systems approach, knowing what mistakes have occurred in the past and integrating hazard identification and control practices to minimize future mistakes.

Business excellence requires what operating excellence produces: trained and motivated personnel, well-maintained and reliable equipment, standardized written procedures that are followed, recognition and management of change, successful performance and safety inspections, a focus on performing each job the safe way, and continuous improvement in all these activities. Business excellence involves, in part, the management of hazards by investigating and setting priorities for hazards' elimination and control.

INDEX